Humanistic Tourism

Human dignity has experienced limited attention in tourism studies. The interlinked dimensions of dignity in tourism urgently ask for broad avenues of future research, as tourism is both an information-intensive industry and an "experience good" resulting from the relationship and co-creation processes involving hosts and guests in different political, socio-economic, cultural, and environmental contexts. These contexts play a role in how an individual's values, norms, and experiences may be experienced in tourism.

This edited book is one of the first attempts to apply to tourism a humanistic management approach entailing a re-discovery of the value of human life, dignity, and awareness of the ethical dimensions of work. The book develops awareness of the contemporary relevance of the human dignity concept to interpret and manage the weaknesses of traditional approaches to tourism and cope with the challenges and new scenarios, including the current COVID-19 pandemic crisis. It presents ethical values and norms as both foundations and vehicles to dignify tourism stakeholders' vision and mission (policy, strategies, and practices) as well as people/tourist beliefs, attitudes, and behaviors. It grounds humanistic education as a pervasive mechanism to innovate tourism management contents and practices by offering to different targets new educational and training formats or framing differently traditional ones. Presenting both a critical and a positive approach to tourism management, the diversity of disciplinary approaches, case studies, and examples make the book attractive to a variety of readers including tourism scholars, researchers, practitioners, and postgraduate students of management and organization disciplines.

Maria Della Lucia is an Associate Professor of Tourism and Business Management at the University of Trento, Italy. Her main areas of research, teaching, and training include local development and sustainability, destination management and governance, culture-led regeneration, creative cities and creative tourism, digital and social media marketing, and economic impact analysis as investment decision-making tools. She has published her research in prominent journals.

Ernestina Giudici is a retired Full Professor of Management and Business Communication at the University of Cagliari, Italy. Her main areas of research include business communication, storytelling and digital storytelling, humanistic management and integrity, corruption, ethical sustainable development, sustainability, and neurotourism. She is part of the managing team of the International Humanistic Management Association and has published widely in various prestigious journals.

Humanistic Management

Series Editors: Michael Pirson, Erica Steckler, David Wasieleski, Benito Teehankee, Ricardo Aguado, and Ernestina Giudici

Humanistic Management draws together the concepts of social business, sustainability, social entrepreneurship, business ethics, conscious capitalism, and cooperative capitalism to present a new humanistically-based research paradigm. This new paradigm challenges the prevailing neo-liberal "economistic" approach that dominates 20th-century management theory and practice and instead emphasizes the need to protect human dignity and wellbeing as well as economic drivers.

Aesthetics, Organization, and Humanistic Management
Edited by Monika Kostera and Cezary Wozniak

Motivation in Organisations
Searching for a Meaningful Work-Life Balance
Manuel Guillén

Humanistic Tourism
Values, Norms, and Dignity
Edited by Maria Della Lucia and Ernestina Giudici

Humanistic Tourism

Values, Norms, and Dignity

Edited by
Maria Della Lucia
and
Ernestina Giudici

Routledge
Taylor & Francis Group
NEW YORK AND LONDON

First published 2021
by Routledge
52 Vanderbilt Avenue, New York, NY 10017

and by Routledge
2 Park Square, Milton Park, Abingdon, Oxon, OX14 4RN

Routledge is an imprint of the Taylor & Francis Group, an informa business

Library of Congress Cataloging-in-Publication Data
A catalog record for this title has been requested

ISBN: 978-0-367-62340-1 (hbk)
ISBN: 978-0-367-62342-5 (pbk)
ISBN: 978-1-003-10899-3 (ebk)

Typeset in Sabon
by MPS Limited, Dehradun

Contents

Figures

Tables

Contributors

Johanna Aalto is a Senior Lecturer at the Laurea University of Applied Sciences in Finland. The main areas of research, teaching, and training include brands, jurisdiction, public management, participation, and local development. Most recently, the focus has been in developing legal value visibility and legal design, how to get legal content more user-friendly and intelligible.

Monica Basile is Marketing & Research Manager for the Italian Alpine Hotel Association, Federalberghi Trentino, in the Dolomites, a UNESCO' Heritage Site; independent Global Tourism Expert at UNWTO, OECD, The European Institute of Cultural Routes. She has 20 years of experience in creating worldwide innovative and profitable hospitality and destination tourism product-experience based on local UNESCO Heritage and on SME businesses activities, suitable for creative tourists. She designed the "Win-Win UNESCO Experience" new training model and she is teaching it as a Guest Lecturer at UNESCO Chair Master's courses in Hospitality, Heritage & Destination Management, also at University Paris 1 Panthéon-Sorbonne, as well as to 3,000 Hoteliers in Executive Trainings and to all Tourism Managers of the Italian Chambers of Commerce. She previously held the positions of Sales and Marketing Country Manager launching worldwide destinations and hotel chains (4,000 "Accor Hotels", and 5* Luxury Hotels of "Sun International" in Southern Africa).

Alfonso Ernesto Benito is an Associate Professor at the University of Monterrey (UDEM). He received his PhD from the University of Deusto (Spain). His thesis focuses on "RSE and Values: Profile from Socially Responsible Companies Based on Brian Hall's Value Model". He is a business consultant and coach and has been an ordinary partner at the European Business Ethics Network–Spain since 2005. He has provided consulting services to companies in the areas of coaching, CSR, and leadership. His research and teaching interests are

CSR, organizational development, human capital, and humanistic management.

Blanca A. Camargo is an Associate Professor of Tourism in the Universidad de Monterrey (UDEM) Business School. She received her PhD in Tourism Management from Texas A&M University. Her research focuses on sustainable tourism, ethics, and justice in tourism planning, development, and management.

Giuseppe Cappiello is an Assistant Professor of Management at the Department of Management of the University of Bologna. He received his PhD in "Strategy, Management, and Quantitative Methods" at the University of Milan, and has been a Visiting Scholar at the Kellog School of Management, Evanston (Il). His research activity is on service management, mainly public services. Current researches focused on intermediaries in innovation, the adoption of artificial intelligence in services, and knowledge management. The latest publication is Cappiello et al. (2020). Social Capital and its effects on networked firm innovation and competitiveness. Industrial Marketing Management. Available online on 18 March 2020.

Mariangela Franch is a Full Professor of Marketing at the University of Trento (Italy). She founded the interdisciplinary Emasus Research Group (Economics, Management, and Sustainable Consumption), which brings competencies in tourism management, marketing, environmental economics, information and communication technologies, and statistics to the analysis of sustainable tourism. She is the coordinator of the Master Program MaST (Sustainability Management and Tourism) and a member of the PhD Program SUSTEEMS (Sustainability: Economics, Environment, Management, and Society) at the University of Trento. She is a member of the International Association of Scientific Experts in Tourism (AIEST). Her main areas of research focus on decision-making processes, behavioral profiles of tourists, and sustainability management. She has authored or co-authored over 180 publications in national and international journals.

Anna Irimias is an Associate Professor in Tourism Management at Corvinus University of Budapest (Hungary). Prior to joining the Corvinus University in 2019, she was teaching in Italy and in Hungary for 10 years. She has been on the teaching staff of the PhD program of Tourism Management at the Corvinus University since 2015. Her most recent co-authored book, Management of Tourism Products, was published by Akadémiai Kiadó in 2020, in Hungary. Her research interests include intercultural aspects of tourism, management of heritage tourism, and film tourism. She has obtained two prestigious post-doctoral scholarships from the Hungarian Academy of Sciences and has been a senior researcher in several national research grants in tourism. She has authored and co-authored more than 110 referred scientific publications in different languages.

Luciana Lazzeretti is a Full Professor of Economics and Management at the Department of Economics and Management of the University of Florence. She is the Scientific Coordinator of the PhD Course of Development Economics and Local Systems (DELOS), University of Florence and Trento, and Director of the Postgraduate Course in Economics and Management of Museums and Cultural Heritage. Her current research interests deal with creative and cultural industries and clusters, creative cities and networks, innovation, resilience and complexity, and artificial intelligence. She has published more than 160 contributions including books, articles, and book chapters. Recently, she published with Vecco M., Creative Industries and Entrepreneurship: Paradigms in Transition from a Global Perspective (2018) and with Cooke P., The Role of Art and Culture for Regional and Urban Resilience (2018).

Arja Lemmetyinen is an Adjunct Professor in international marketing and currently working as a University Researcher (marketing) at the University of Turku, School of Economics, Pori Unit, Finland. Her research interest lies first in brand management and place branding. She is also interested in research on network management and value co-creation in networks. Lately her research interest has focused on sustainability in its economic, social, environmental, and cultural significance, and also in digital brand experiences. She has published her research widely in Journal of Brand Management, Place Branding and Public Diplomacy, Marketing Intelligence & Planning, Tourism Management, Tourism Review, and Scandinavian Journal of Tourism and Hospitality. She serves, among others, on the editorial board of two learned journals, Journal of Enterprising Communities: People and Places in the Global Economy, and the International Review on Public and Nonprofit Marketing.

Plinio Limata is a Postdoctoral Research Fellow on "Platform economy" and "blockchain" at LUMSA University of Rome. He holds a PhD in Civil Economy and is an expert in "Economic Policy" and "History of the trade union movement". He graduated with honors in political sciences of international relations, and holds a master's degree in "Experts in politics", and one in "Social Innovation and Social Business". He has worked as the Project Manager responsible for Africa and South America for the NGO ISCOS, and he was the Representative of the Italian Trade Union Confederations at the 107th International Labor Conference promoted by the ILO. He is also a Founding Member and Chairman of the Board of Directors of the

"International Napoli Network" cooperative and founding member of "Impact Hub Roma Srl". He has experience of study and work in the field of international and diplomatic relations, as well as international cooperation and volunteering between the USA (NY), Saudi Arabia, Ecuador, Brazil, and Mozambique.

Adele Magnelli is an International Project Manager in ETT S.p.A. and an expert in multimedia exhibitions management. In New Media since 2008, she strongly believes in the added value of new technologies in the cultural sector to transform familiarity into knowledge. She combines creative ideas, innovative design, and storytelling with state-of-the-art technologies to generate immersive experiences for museums and corporate spaces. Her projects include the mixed reality experience "The Ara as It Was" in Rome, "Ars Excavandi" for Matera 2019 opening exhibition, and the immersive experience of the Vicinato a Pozzo, in Matera. She also managed many projects in the luxury field for exclusive clients, including Bulgari. In the academic field, she holds lectures about ICT in Cultural Heritage and Tourism, and storytelling. She has published several papers, including the 2019 Best Conference Paper at the 5th International Augmented and Virtual Reality Conference in Munich.

Ariel Mitev has been Associate Professor in Marketing at Corvinus University of Budapest (Hungary) since 2006. He received the Excellent Mentor Award in 2019. He has been on the teaching staff of the PhD program of Marketing and Management at the Corvinus University since 2010. His interests include marketing research methodology, scale development, consumer behavior, and social marketing. He has widely published on marketing issues in referred journals authoring and coauthoring over 120 publications in national and international journals, including several books on qualitative and quantitative research methodology.

Gabriele Morandin is an Associate Professor of Organizational Behavior at the Department of Management of the University of Bologna and Associate Dean at Bologna Business School, Italy. His research is focused on employability and its effects over time and over contexts. His publications appeared among others on the Computers in Human Behavior, Employee Relations, European Journal of Marketing, European Management Review, Harvard Business Review, Journal of Applied Psychology, Journal of Management Inquiry, MIT Sloan Management Review, and Scandinavian Journal of Management, while he seats on the Editorial Board of the Journal of Vocational Behavior.

Lenita Nieminen is currently working as a Project Researcher (Entrepreneurship) at the Turku School of Economics at the

University of Turku. She has a 20-year background in creative economies. Her research focuses on entrepreneurial learning and the role of entrepreneurial networks in micro-business development. She has been engaged as a researcher in several regional and international entrepreneurship projects, specializing in cultural tourism. She has co-authored articles published in the International Journal of Entrepreneurship and Innovation, the Journal of Entrepreneurship + Training, the Journal of Enterprising Communities:People and Places in the Global Economy, Innovation, and Entrepreneurship in Education (Advances in Digital Education and Lifelong Learning).

Giorgia Nigri is a practitioner and active researcher in the field of economics. She works for Italy's export credit agency since 2013 and is a certified benefit impact manager. She is a member of various alumni associations, as she has been a business consultant for over 10 years and collaborates with researchers in several other disciplines through academic networks. Giorgia completed her PhD at LUMSA University, her MBA at the Luiss Guido Carli Business School, and her undergraduate studies at John Cabot University and University of Wales. Her research interests lie in the area of sustainability, sustainable development, and corporate social responsibility with attention to new hybrid forms of business such as benefit corporations and social purpose companies.

Stefania Oliva is a Post-doc Fellow researcher at the Department of Economics and Management, University of Florence. Her research interests focus on the resilience and sustainability of cities and regions, cultural organizations, and creative industries. She studies culture and creativity as sources for adaptation and adaptability of regions to shocks. She is interested in the gender issue, art cities and fashion cities, social and environmental reporting and accountability in museums. She has recently published in European Planning Studies, City, Culture and Society, Papers in Evolutionary Economic Geography, among others.

Sabina Owsianowska is an Assistant Professor at the Faculty of Tourism and Leisure, University of Physical Education in Krakow; and cooperates with the Institute of Intercultural Studies, Jagiellonian University (Poland). Her main areas of research, teaching, and training include the theory of tourism and leisure; anthropology of tourism; tourism education and management in a humanistic perspective; multiculturality and the interpretation of heritage, with focus on its dissonance, mainly in Central and Eastern Europe; visuality of tourism and the embodiment of travel experiences; and qualitative methods. Recently, she authored and co-authored three monographs (e.g., Anthropology of Tourism in Central and Eastern

Europe. Bridging Worlds, Rowman & Littlefield, 2018, co-edited with M. Banaszkiewicz), several thematic issues of tourism journals (e.g. Journal of Tourism and Cultural Change 15/2017), and numerous papers and chapters. She is a member of the scientific board of Folia Turistica, and the teaching staff at the Master program in Tourism and Leisure, and Intercultural Relations in Tourism.

Manuela Presutti is an Associate Professor of Strategy and Management at the Department of Management of the University of Bologna and Director of Second Cycle Degree of Tourism Economics and Management at the Campus of Rimini. Her main research activity was originally focused on small firms, entrepreneurship, and the internationalization process. During the last few years, her research interest has included the analysis of social networks, the ties between social capital and new firms creation, industrial clusters, and the tourism industry. She has published on these topics, in national and international journals, such as International Business Review, Entrepreneurship, and Regional Development, Journal of Small Business Management, and Journal of World Business.

Michael D. Santonino III is an Associate Professor of Management and Marketing in the Department of Business Administration at Embry-Riddle Aeronautical University-Worldwide. He has developed and taught management and marketing courses at the graduate and undergraduate level for ERAU, Bethune-Cookman University, and Daytona State College, all with main campuses in Florida. He earned his DBA & MBA at the H. Wayne Huizenga School of Business and Entrepreneurship from Nova Southeastern University and holds an Electrical Engineering Degree from the New York Institute of Technology. Dr. Santonino is a former member of the technical staff at AT&T Bell Laboratories/Lucent Technologies in Orlando, Florida, and was a Senior Design Engineer at Motorola in the Land Mobile Products Division in Ft. Lauderdale, Florida. His research interests include management, inclusive organizations and sustainability marketing. He is the co-author of The Supervisor's Guide to Human Relations and Communications.

Davide Secchi is an Associate Professor of Organizational Cognition at the Department of Language and Communication, Director of the Research Centre for Computational & Organisational Cognition, University of Southern Denmark, Slagelse. His research is organized around (a) rationality in distributed cognitive environments by using (b) empirical data and advanced computational simulation techniques, especially agent-based modeling. Long term research interests also include the theory and practice of individual and corporate social responsibility. Recent research has been directed towards re-defining organizational cognition through change, with particular reference to plasticity and disorganization. He is the author of 32 journal articles, 21 book chapters, 4 between edited books and monographs, and 40 various other publications. Chair of the track "Research Methods for Complex Adaptive Systems" at the European Academy of Management annual conference, he is the convenor of the Agent-Based Models of Organizational Behavior Workshop Series and has presented his work more than 100 times at major international conferences (2002-2020). Davide sits on the editorial board of Kybernetes, Team Performance Management, and, since March 2020, he is also Editor-in-Chief of the International Journal of Organization Theory & Behavior (Emerald).

Mario Vázquez Maguirre is an Associate Professor of Management and Social Entrepreneurship in the Universidad de Monterrey (UDEM). He received his MSc in Economics and Public Policy from Tecnológico de Monterrey and his PhD in Management from EGADE Business School. His research focuses on social entrepreneurship, indigenous social enterprises and sustainable development, humanistic management, and corporate social responsibility. He has documented cases of Indigenous Social Enterprises in Latin America and published his findings in leading journals and book chapters.

Atila Yüksel is a Professor of Services Marketing at the University of Aydın Adnan Menderes University (Turkey). Main areas of research, teaching, training include andragogy in tourism education, tourism planning, tourist behavior, co-creation, destination marketing and branding, research methodology, talent management, and travel blogging. Within his areas of interest, he authored and co-authored books on tourist satisfaction, research methodology, and cultural heritage management, together with numerous book chapters and five best-selling novels. He has widely published in leading tourism journals and listed as one of the most prolific scientists in Webometrics Turkey. He has been acting as a board member in numerous journals and he is the co-founder of Travel and Hotel Management Journal, which has been honored as one of the leading travel and tourism journals in Turkey. He teaches as a visiting lecturer at Istanbul Bilgi University and Ozyegin University, and gives consultancy services to European Union projects of Ministries and Chambers of Commerce.

Preface

Maria Della Lucia and Ernestina Giudici

The edited book *Humanistic Tourism: Values, Norms, and Dignity* is one of the first attempts to apply a humanistic management approach to tourism. This entails a rediscovery of values such as human life, dignity, meaningful work, and the ethical dimensions of working. The book is part of a book series that includes the edited book *Humanistic Management and Sustainable Tourism: Human, Social and Environmental Challenges*, which stands at the nexus between tourism and destination management, and humanistic management. Tourism and destination management is an area of study that has attracted attention in the recent past and has grown significantly, becoming an established business discipline. Humanistic management has been the subject of increased research in the last few decades, and is on the rise, although it is still relatively new. Integrating the two approaches offers a novel and original perspective on tourism that has not yet attracted much research.

This intersection is promising from both theoretical and managerial perspectives. The topic of human dignity, so far, has been given limited attention in tourism studies (Winchenbach, Hanna, & Miller, 2019). While the subject of accounting for human dignity features prominently in many tourism-related national and international policies (Transforming Tourism, 2017; UN, 1948; UNWTO, 2001), a review of tourism academic literature shows that, to date, the contributions of academia have been few. Even those have mostly been recent and had a narrow focus. The reason is that the subject of human dignity is a powerful yet complex concept, which takes on a variety of meanings. Many tourism studies provide some insight into the subject only in passing, while investigating other issues, such as corporate social responsibility and sustainability, among others.

The interlinked micro, meso, and macro levels of human dignity in tourism offer broad avenues for urgently needed research. Tourism is both an information-intensive industry and an "experience good", resulting from the relationship and co-creation processes involving hosts and guests (Gretzel, Fesenmaier, Formica, & O'Leary, 2006), in different political, socio-economic, cultural, ethnic, and environmental contexts. These contexts have a role in how dignity, values, and norms are experienced in tourism (Baum, 2006; Sayer, 2011), affecting the development and professionalization of humanistic tourism research, practice, policy, and pedagogy.

By adopting a humanistic management perspective, the book covers several aspects of tourism management. It presents a number of disciplinary approaches from a critical perspective. It discusses case studies and examples that cover a wide variety of contexts, providing variations and applications of disciplines, concepts, and ethical values. It also offers new educational and training formats or frames differently traditional ones. For these reasons, the book is a resource for a variety of audiences – tourism scholars, researchers, and graduate students of many sub-disciplines of management and organization research, along with practitioners interested in gathering information and fresh knowledge regarding humanistic tourism. This promising area of study, teaching, practice, and education sheds some light on the potential and importance of adopting a humanistic management perspective and implementing it more widely in tourism.

The edited volume aims at the following:

- Developing an awareness of the contemporary relevance of the concept of dignity to manage the weaknesses of traditional approaches to tourism;
- Presenting ethical values and norms as both foundations and vehicles to bolster the role of dignity in tourism management – in tourism's stakeholders' vision and mission (policy, strategies, practices) and in the behaviors, attitudes, and beliefs of both people and tourists; and
- Highlighting how humanistic education acts as a pervasive mechanism to innovate tourism management contents and practices at different levels (education formats, professional and business training models, educational experiences).

Book Contents and Structure

The edited book contains thirteen contributions, organized into two separate parts, and a conclusion section. They analyze and discuss selected issues of relevance in tourism management theory and practice, both from the supply (governments and business) and demand perspective (the tourist). These issues are introduced by the editors in the opening chapter of the volume *Humanistic Tourism: Exploring Old and New Tourism Challenges from a Humanistic Perspective* (1). Maria Della Lucia and Ernestina Giudici set humanistic tourism at the nexus between tourism management and humanistic management, striving to establish a new disciplinary perspective on tourism, which can interpret and manage the weaknesses of traditional approaches to tourism and cope with tourism challenges and new scenarios, including the current global pandemic crisis. Putting human subjects at the center should help institutions and organizations question the presumed rationality of the homo economicus and lever on dignity, justice, and equity to promote fair practices in employment, value creation, delivery and capture in economy and society. The issues raised cluster around the underlying question of this book, entailing a paradigm shift in tourism: Is it urgent, mandatory, and feasible or still a utopia? The debate around the current global pandemic crisis and its effects on tourism shows that the answer is complex and has both bright and dark sides.

Far from providing full guidelines, the book draws attention to some theoretical contributions and case studies that are aimed to inspire and help greatly to understand and grasp the relevance and importance of adopting a humanistic management perspective in tourism to foster a paradigm shift.

Part I "Criticism and Ethical Value in Tourism" incorporates five chapters sharing a theoretical focus on and interest in humanistic tourism across different contexts which affect (ethical) value in tourism and has different areas of tension.

In 2, "The Truths About Conventional Tourism and Its Scientism: A Humanistic Critique", Atila Yuksel focuses on the predominant neoliberal structure of tourism, which has both pros and cons. The emphasis on the economic role of tourism – its impact on employment, income creation, infrastructure, business and service development, as well as tourist flow generation – produces also a set of beliefs, attitudes, and measures, which legitimate how tourism should develop and how its research ought to be. The chapter undertakes a humanistic critique of the inherent costs and problems of this self-reinforcing cycle.

Written by Plinio Limata and Giorgia Nigri, 3, "Voluntourism, Humanistic Management, and Relational Goods" frames voluntourism as a growing topic of interest for both academics and practitioners which has led to conflicting judgments, both positive and negative. The authors provide a new interpretative framework for voluntourism in light of the theory of the relational good and under the umbrella of humanistic management, pointing out the role of this form of niche tourism in helping volunteers and hosting communities to fulfill their potential by building bridges in order to connect the value embedded in both parties.

In 4, "Humanistic Tourism through the Mayan Train Megaproject: A New Narrative for Tourism Development in the Yucatán Peninsula", Blanca A. Camargo, Mario Vázquez-Maguirre, and Alfonso Ernesto Benito address the neglected issues of human dignity, well-being, and human flourishing experienced by indigenous groups in relation to the impact of uncontrolled and unplanned tourism development on their natural habitats and their cultural fabric. Drawing upon humanistic management, the Authors propose a new narrative to protect and promote human dignity in tourism development by discussing the case of the Mayan Train mega-project in the Yucatán Peninsula of Mexico. This narrative pays particular attention to the potential and the responsibility of government and private industry in recognizing the human dignity of Mayan communities. It levers on five general principles to promote compassion, happiness, and human dignity in order to achieve human flourishing and well-being through tourism.

In 5, "Ethical Value Propositions in Branding a Rural Community", Arja Lemmetyinen, Lenita Nieminen, and Johanna Aalto examine how a humanistic management view of marketing may be reflected in place branding with the offering of ethical value propositions. By so doing, the research builds on the service-dominant logic dominating contemporary marketing discourse, and the identity-based place branding, emphasizing the dialogue and co-creation between place branding stakeholders. The ethical value propositions of stakeholders collaborating in place branding are examined in a rural community in Finland. The value propositions co-created by rural stakeholders clearly stem from the culture of the place interwoven in its image and identity. This insight implicitly covers the humanistic view of management, with its stakeholder focus.

The last chapter (6) of Part I, "Service Delivery and Employee Turnover: The Importance of Work-Life" written by Giuseppe Cappiello, Gabriele Morandin, and Manuela Presutti, has a micro perspective on humanistic tourism. In a labor-intensive service industry such as tourism, human resources are essential to delivering services and satisfying customers, however, staff turnover in tourism businesses is very high. Study results indicate that commitment and job satisfaction influence the intention to remain with the same company over time and organizational identification has a significant effect on commitment. On the other hand, work-life balance and the quality of the relationships with managers are among the main causes behind a change in jobs. These factors are at the core of human dignity in the workplace.

Part II, "Innovating Education, Training, and Experiences in Tourism" incorporates six chapters focusing on humanistic education and training as crucial activities to innovate management practices and contents, aiming to cope with the areas of tension in tourism management and to enhance the educational tourism experience. New educational and training formats, or traditional ones framed differently, are provided.

7, "Tourism Education from a Humanistic Perspective: Anthropological Inspirations" written by Sabina Owsianowska, focuses on anthropological knowledge as a crucial competence for tourism practitioners. Managers, hotel employees, tour guides, heritage interpreters, cultural brokers, etc., all deal with such issues as encounters of tourists with other people, cultural conflicts and misunderstandings, host-guest relationships, or the dissonance of heritage. Tourism staff not only provide services but, above all, they create, promote, and share worldviews, values, and meanings. They shape tourist encounters and practices and may contribute to the protection of human dignity, ontological security, traditional lifestyles, and endangered heritages, which is why their knowledge, awareness, sensitivity, and 'soft' skills matter.

In 8, "Humanistic Management Training: An Explorative Study on Erasmus Students' Intercultural Sensitivity", Anna Irimiás, Mariangela Franch, and Ariel Mitev focus on the development of the intercultural sensitivity of students, as future managers and tourism practitioners working in a global market. It has to do with an empathetic, ethical, and non-discriminative mindset, which is at the core of humanistic management. University education programs are rarely aware of the importance of establishing courses, which address intercultural sensitivity and competences. An analysis exploring whether and how Erasmus study abroad programs influence tourism management students' critical consciousness about cultural diversity reveals that students perceive themselves as ethnorelative, considering different worldviews a value, something to respect, and to endorse. Universities should provide training sessions on intercultural sensitivity, both before departure and after the students' return from abroad, in order to stimulate reflection on the experience and the development of self-identity as well as managerial skills.

In 9, "A New Entrepreneurial Training Model: The "Win-Win UNESCO Experience", Monica Basile presents a capacity-building model aimed at involving destination stakeholders in the creation of an experiential tourism product. It consists of the joint public-private local stakeholders capacity-building modules tailored specifically for UNESCO destinations. While creating a product that attracts new segments of culture demand, this training model allows local entrepreneurs to reinvent their social role within the community. They lever on intangible heritage and local know-how to co-create value, enhance destination well-being, and engage tourists in local experiences.

In 10, "The Humanistic Supervisor", Michael Santonino focuses on the humanistic supervisor meant as a manager who is sociable, community-aware, and well-informed about the promotion of holistic human well-being. Data collected during observational field trips in Singapore and Florence, Italy (both UNESCO heritage sites), which include images, spontaneous conversations, and unstructured interviews, offer practical insights into the challenges facing humanistic supervisors. Their priorities should include a shared interest for the common good in tourism employment and a desire to educate people about sustainability in ways that might help them to change their behaviors and work for the common good to benefit society, businesses, and the planet. The chapter offers ways to train future tourism managers, increase stakeholder participation, and encourage grassroots initiatives to (gradually) change the economy-driven tourism industry.

11, “Natural History Museums and Sustainable Development: The Role of Education for Humanistic Tourism”, written by Stefania Oliva and Luciana Lazzeretti, reframes the traditional educational role of museums in the humanistic perspective. Mapping and describing the activities (carried out by the Museums of Natural History of Florence, Italy), for disseminating knowledge of science and nature, the analysis shows that it may represent an informal space where to develop participatory forms of visitors’ education on topics related to environmental sustainability. Results suggest that the educational function of the museum involves events and exhibitions for the engagement of public and audience development. Moreover, the museum’s research activities and scientific publications contribute effectively to safeguarding biodiversity and enhancing naturalistic collections. Humanistic tourism includes the development of this knowledge and sensitivity.

In the last chapter (12) of Part II “How Museums can Contribute to Collective Well-Being: The Role of an “Augmented” Video Guide”, Adele Magnelli explores the connection between cultural heritage museums fruition and psychological well-being. Multimedia museums, as the case of the *Musei Civici di Roma Capitale*, can be the means to reach well-being thanks to state-of-the-art technologies such as Augmented Reality and Virtual Reality. Humanistic management, rooted in the great humanistic and Renaissance heritage of Italian and European culture, is part of this discourse. It uses tools, such as the different forms of art and cultural heritage unlocked by new digital and multimedia technologies, to overcome sectorial languages and provide immersive educational experiences to different audiences.

In the concluding section (13) “Reshaping Tourism: Advances and Open Issues”, Maria Della Lucia, Ernestina Giudici, and Davide Secchi discuss the practical application of humanistic management to tourism and new avenues of research. The dimensions of humanistic tourism, the humanistic difference, and the humanistic advantage allow the extending to tourism organizations and their environments of the complex adaptive systems perspective. This approach reinterprets human dignity, values, norms, and education through the lenses of uncertainty, heterogeneity, and context. It also paves the way for further developing humanistic management through sustainability and social responsibility, a specific area of inquiry that naturally connects to tourism as a complex system.

References

Baum, T. (2006). *Human resource management for tourism, hospitality and leisure: An international perspective*. London: Thomson Learning.

Gretzel, U., Fesenmaier, D. R., Formica, S., & O'Leary, J. T. (2006). Searching for the future: Challenges faced by destination marketing organizations. *Journal of Travel Research*, *45*(4), 116–126.

Sayer, A. (2011). *Why things matter to people: Social science, values and ethical life*. Cambridge: Cambridge University Press.

Transforming Tourism. (2017). *Berlin declaration on "transforming tourism"*. Retrieved from http://www.transforming-tourism.org/berlin-declaration-on-transforming-tourism.html.

United Nations (UN). (1948). *Universal declaration of human rights*. New York: United Nations.

Winchenbach, A., Hanna, P., & Graham, M. (2019). Rethinking decent work: The value of dignity in tourism employment. *Journal of Sustainable Tourism*, *27*(7), 1026–1043. https://doi.org/10.1080/09669582.2019.1566346.

World Tourism Organization (UNWTO). (2001). *Global code of ethics for tourism*. Madrid: World Tourism Organization.

Acknowledgments

The editors are especially grateful to all the authors who, despite competing commitments, made this collaborative project possible. We also appreciate very much their cooperation in such a positive manner. Also, many thanks to the reviewers for their valuable and constructive critiques: Anke Winchenbach (University of Surrey, United Kingdom), Anna Scuttari (Munich University, Germany and EURAC Research, Bozen/Italy), Emanuela Conti (University of Urbino, Italy), Fabio Forlani (University of Perugia, Italy), Giacomo Del Chiappa (University of Sassari, Italy), Linda Osti (University of Bozen, Italy), Michela Floris (University of Cagliari, Italy), Oksana Tokarchuk (University of Bozen, Italy), Tom Griffin (Ryerson University, Toronto/Canada), Volker Rundshagen (Stralsund University, Germany), Werner Gronau (Stralsund University, Germany). The editors and authors also express their appreciation to the rights holders of some of the illustrations used in this book, for their permission to reprint this material. Finally, we gratefully acknowledge Michael Pirson (Fordham University, New York), Co-Founder and President of the International Humanistic Management Association and Editor of the Humanistic Management Series at Routledge, as well as the Routledge publishing staff, in particular Brianna Ascher and Naomi Round Cahalin, who were ever-present and supportive.

1 Humanistic Tourism

Exploring Old and New Tourism Challenges from a Humanistic Perspective

Maria Della Lucia and Ernestina Giudici

1 Introduction

Tourism is a social, cultural, and economic phenomenon that entails the movement of people to countries or places outside their usual environments for personal, business, or professional purposes. This well-known definition of tourism of the United Nations World Tourism Organization captures a phenomenon that, involving the movement of human beings from one place to another and their encounters, constitutes a transformative force of space, places, and people.

The exponential growth of tourist flows has given rise to mass tourism, a form of tourism that started to develop in the second half of the last century and has increasingly been driven, in a dramatic fashion, by globalized, consumer-driven societies and key transformations in technology and mobility. The postmodern era has also experienced the rise of – or the return to – niche forms of tourism, which cannot be seen as the long tail of mass tourism but as a wide range of small-scale segments that have different motivations and that seek different places and offers (Clarke, 1997).

The coexistence of old and new forms of tourism (Della Lucia, 2018) reveals tensions and opportunities that challenge tourism and destination management theory and practice. The rampant phenomenon of overtourism is one symptom of the exploitation and damages generated by an excess of tourism over a destination's capacity, as evidenced in a wide range of locations (Dodds & Butler, 2019), leading to conflict with and complaints by residents concerned that overtourism is making their homes unlivable and environmentally unsustainable. Conversely, niche tourism occurs in the context of complex individual personalities and evolving motivations, transforming tourism into meaningful and learning patterns defined for, and with the participation of, tourists in sustainable destinations (Della Lucia, 2018).

This chapter positions humanistic tourism at the nexus between tourism management and humanistic management, aiming to establish a new disciplinary perspective on tourism that can interpret and manage the weaknesses of traditional approaches to tourism and cope with tourism challenges and new scenarios, including the current global pandemic crisis. In so doing, this chapter covers the evolution of tourism, pointing out its main challenges. By focusing on the recognition and protection of the value of people, justice, and equity, humanistic tourism promotes fair practices in employment, value creation, delivery, and capture within organizations and societies. The underlying issue is a paradigm shift in tourism that is urgent but complex and still open.

2 Tourism's Evolution

Tourism has very ancient roots dating back to the Greek and Roman civilizations and has undergone multiple transformations over the centuries. Butler (2015) noted:

> Throughout its history, tourism has been characterized by both inertia and dynamism. Many of the old patterns of travel and behavior still exist, albeit modified by technological advancement, reflecting human preferences for constancy and a dislike of change, while new destinations, new attractions, new methods of travel, and reduced costs have all enabled both a vast increase in tourist numbers and a more varied selection of activities to be engaged in (p. 5).

After the Second World War, many states had to prioritize reconstruction, leading to an economic reawakening that has influenced the life, habits, and sociality of an increasing number of people (Pedersen, 2020). At the same time, economies have become increasingly open. Technological development has favored the rapid movement of goods and people. Tourism has played a primary role in the economic development of many areas of the West and has also become one of the main drivers of growth in developing countries.

Due to the size and growth of international tourism flows and their economic impact, tourism is normally referred to as the largest industry in the world. More precisely, tourism is the world's largest service industry in terms of international trade (Lew, 2011). Its size and impact depend on how tourism is defined. This industry spans different sectors – which may not always be classified or classifiable as tourism – that directly or indirectly produce goods and services consumed by both tourists and residents, as well as by tourism companies and the public sector.

Driven by the pro-growth approach, the exponential and pervasive trajectory of tourism development emphasizes profitability, economic well-being, and employment and consumption opportunities, as well as the creation and attraction of businesses, human resources, and investments. The marketization and corporatization of tourism have served the interests of the powerful players in the tourism industry (Higgins-Desbiolles, 2006) at environmental, social, and ethical costs (Lee, Lee, Choi, Yoon, & Hart, 2014). Due to the progressive globalization of tourism, the costs of this (seemingly) limitless growth have been paid in developed and developing countries alike (Ajagunna, 2014). Local communities have been excluded from participation in decisions about economic opportunities, culture, and relationships with the natural environment and marginalized in the distribution of resources for tourism development (González Tirados, 2011).

These dynamics have triggered a debate about (sustainable) tourism planning and management (Inskeep, 1991; Swarbrooke, 1999). Tourism development has to balance the economic impact with the conservation of the irreproducible resources involved in tourism production, in particular the natural resources and the cultural heritage of the host community. The restoration of rivers and surrounding areas in various parts of the world (e.g., Singapore; San Antonio, Texas; and Cheonggyecheon, Seoul) indicates how some areas have "incorporated fundamental ideas of sustainability by combining traditional cityscapes with modern commercial activities" (Lee et al., 2014, p. 227). The balance of economic, environmental, and sociocultural goals also has to do with a return to the view of tourism as a social force aimed at achieving important social goals (Higgins-Desbiolles, 2006), including equity, justice, fairness, and social change.

On the demand side, this approach entails the formulation of strategies addressing growing segments of sustainable demand, whose motivations cluster around the four main dimensions (4Ls) of experience (Franch, Martini, Buffa, & Parisi, 2008). Leisure consists of both hedonic and eudaimonic forms of experience encompassing pleasure but also meaningful self-help and self-development activities as ways to fulfill human potential (Parsons, Houge Mackenzie, & Filep, 2019; Smith & Diekmann, 2017). Landscape involves the interest in admiring, moving in, and safeguarding nature through nature-based tourism and tourism in protected areas (Hockings et al., 2020). Learning has to do with discovering, experiencing, participating in, learning about, and becoming more intimately involved in the everyday life of people and places visited (Russo & Richards, 2016). Limits are associated with the limited amount of local resources, as well as the recognition and respect of the value of people and nature in awareness of the social and environmental impact of experience, or responsible practices in tourism (Higgins-Desbiolles, 2020). This internationally increasing demand has a great potential as it combines sustainable motivation with consistency – but not mass flows – and tourism, along with the environmental and social limits of the impact of tourism (Della Lucia & Martini, 2012). The transition to greener, slower, more resilient tourism remains a challenge.

3 Attacks on Tourism Growth

Tourism growth trends have ground to a halt several times over the decades due to external factors (e.g., World Wars I and II), terrorist attacks (e.g., the attack on the World Trade Centers in New York on September 11, 2001), and pandemics such as SARS and bovine spongiform encephalopathy (BSE), the West Nile virus, and the current COVID-19 pandemic (Gössling, Scott, & Hall, 2020).

These dramatic events have had a common denominator: the slowdown or cessation of tourism (temporarily), the damage of economies and economic well-being, the reaction to tourism degrowth through recovery strategies, and behaviors aimed at returning to tourism as usual. However, these events may also represent an opportunity for beginning tourism again on new grounds.

Milano, Novelli, and Cheer (2019) pointed out the need to rethink tourism by limiting its continuous expansion or by guiding the tourism industry toward a (relatively) smooth degrowth transition. Growth is not equivalent of development. Liberal policymakers in the 1950s and 1960s, as well as some scholars (Chandler, 1963), considered the dimensional growth of companies as the essential matrix for economic development; however, reality has highlighted the erroneousness of this correlation. Above all, firms' growth is not equivalent of sustainable development, the latter being an alternative to unlimited growth (Hall, 2010; Novelli, 2015). Studies on degrowth have confirmed this notion (Latouche, 2014; Schneider, Martinez-Alier, & Kallis, 2011) and questioned the hypotheses on which the continuous expansion of tourism is based (Higgins-Desbiolles, Carnicelli, Krolikowski, Wijesinghe, & Boluk, 2019). Hall conceived of degrowth as occurring within sustainable limits, "not so much connected to downsizing per se but to 'right-sizing'" (Hall, 2009, p. 55), aiming at limiting the negative environmental and social impacts of (unlimited) tourism growth.

In this vein, the call for tourism "degrowth" has gained increasing support as a way to deal with overtourism in different areas as an emergency aspect in tourism studies (Andriotis, 2018; Hall, 2009). Degrowth constitutes a strategy to limit flows within the destination's environmental and social carrying capacity. To be effective, degrowth strategies must be progressive, inclusive, and fair: Degrowth calls for a voluntary, planned contraction rather than haphazard reactions that could exacerbate the phenomenon. It is a matter of moving from "growth for development" to "degrowth for liveability" (Milano et al., 2019).

The innovative aspect of the debate concerning the problems associated with overtourism and the call for degrowth is bottom-up leadership. Supporters, activists, and civil society groups who live in contexts burdened by an excess of tourism seek to reduce the impact that overtourism has on the individual, social, and environmental spheres.

These insights are strictly connected with the principles making sustainable tourism effective – planning and wide stakeholder participation in decision-making (Godfrey & Clarke, 2000). Strategic planning frames tourism in the wider process of territorial governance, and it is cross-sectoral and spans diverse policy domains, thus requiring the coordination of policies in different sectors to create sustainable tourism (Bramwell & Lane, 2011; Hall, 2008). The adoption of a participatory approach to tourism planning can also develop sustainability awareness and encourage the sharing of responsibility for alternative management strategies.

Kallis, Kostakis, Lange, Muraca, Paulson, and Schmelzer (2018) argued that it is possible to consider reducing tourism supply and demand without reducing prosperity and well-being. Moreover, Kallis (2011) stressed that "sustainable degrowth is not only an inevitable hypothesis but also a potent political vision that can be socially transformative" (p. 873). Regarding whether degrowth makes tourism reshaping indispensable, Higgins-Desbiolles (2018) offered a new notion of tourism as "the process of local communities inviting, receiving, and hosting visitors in their local community, for limited time durations, with the intention of receiving benefits from such actions" (p. 12).

This definition does not apply in every context, but it is consistent with what is happening in a growing number of small areas worldwide. Such areas include Guna Yala, Panama (Higgins-Desbiolles et al., 2019) and Sfantu Gheorghe, Romania, in the Danube Delta (Ivan, 2017), as well as many parks, protected areas, and natural World Heritage Sites (Della Lucia & Franch, 2017; Dettori, Caboni, & Giudici, 2017; Giudici, Dettori, & Melis, 2012). In these successful cases, the community is actively involved in choices regarding its place, values, and identity. "Livable" tourism requires people to put the inherent value of human beings and their well-being at the core. In this perspective, the forms of tourism presented as "inclusive" should not involve local realities in a nominal or non-substantial way (Higgins-Desbiolles, 2018). This principle should be embedded in the degrowth of global tourism and the transition to new models of tourism to address existing challenges and current and future crises.

4 Humanistic Management

Humanistic management seeks a change in the current economic paradigm by focusing on the unconditional human dignity of every individual (Spitzeck, 2011). The economistic management approaches (Hamel, 2007) – based, among the *other*, on Taylor's scientific management (1911), Fayol's functions of management (1917), and Elton Mayo and colleagues' refocus on managerial strategy incorporating the socio-psychological aspects of human behavior in organizations – were all aimed at improving organizations' economic efficiency, especially labor productivity. While guiding organizational success, the principles shaping the classical management paradigm (i.e., planning, organizing, leading, and controlling; Scott, 2003) did not sufficiently address crucial organizational issues such as the recognition and protection of human value, the promotion of human flourishing and ethical development, or social justice and social welfare.

This void has emerged and showed weaknesses in managing organizations (Hitt, 2000) in light of the unpredictable changes in markets, technology, and demand (Burnes, 2005; Secchi, 2011). The new complexity in this industry has cast doubt on the efficacy of the classical management paradigm and its obsessive attention to profit and its maximization (Secchi & Newmann, 2015). It has also prompted a shift to new paradigms (Pirson & Lawrence, 2009) that can reply to the increasing need for flexibility and dynamicity in managing companies in complex scenarios (Bocchi & Ceruti, 2007).

The remarkable versatility of human beings (Melé, 2003) is at the core of this new approach, recognizing that the "human being is both rational and emotional, has talents and creativity, and therefore the potential to introduce (business) innovation" (Melé, 2016, p. 38). In so doing, *Homo economicus* is replaced by *Homo sapiens* at the core of the humanistic management perspective.

Melé (2003) defined humanistic management as "a management that emphasizes the human condition and is oriented to the development of human virtue, in all its forms, to its fullest extent" (p. 79). Humanistic management has developed over the decades including different facets, which Melé (2016) documented, such as the evolution from an alternative to "mechanistic" management, to the humanizing of business, to the humanism in economies and societies.

The need for the humanization of businesses, the economy, and society appears evident considering one of the pillars of humanistic management: considering businesses (and organizations in general) not as mere "communities" but as "communities of persons." A "community of persons" is a concept of great human richness and also comprehensive of the complexity of the links which maintain people united within a business enterprise" (Melé, 2012, p. 9).

The concept of humanistic management has thus evolved in scope and space: Fostering unconditional human dignity is the ultimate goal of an organization, and it is a necessary condition for archiving well-being in the organization and prosperity in the economy and society (Pirson, Martin, & Parmar, 2017). In this sense, humanistic management may provide a framework that guides managerial practices toward equity, justice, and fairer outcomes for humanity by interconnecting human beings, businesses, and their social and environmental contexts. Humanistic management may also contribute to remedying the injustice and unfair access to basic resources emerging in different economic and social dynamics.

5 Humanistic Tourism: A Novel Perspective

The growth of international tourism and its environmental and social vulnerabilities offer a practical perspective of humanistic management at both the business and destination levels and from both the demand and supply sides.

Humanistic tourism may be set at the nexus between tourism, destination, and humanistic management to establish a new disciplinary perspective on tourism. Such a new approach could allow scholars to interpret and manage the weaknesses of traditional approaches to tourism and cope with tourism challenges and new scenarios, including the current COVID-19 pandemic. This novel and original perspective on tourism has not yet attracted much research and is promising from both theoretical and managerial viewpoints.

The tourism literature has grown significantly and become an established business discipline. However, researchers have paid little attention to human dignity. A review of the tourism literature suggests that, to date, the academic contributions are few and recent, and they have a narrow focus (e.g., on decent work, dignity, and indigenous justice; Camargo & Vázquez-Maguirre, 2020; Winchenbach, Hanna, & Miller, 2019). Research on this topic is lacking because dignity is a powerful yet complex concept that assumes multivalent meanings. Furthermore, many tourism studies implicitly provide insights into dignity when investigating other issues, including corporate social responsibility and sustainability.

Nevertheless, many national and international tourism-related policies address dignity violations across the sector and beyond, calling on people to respect workers' dignity and to meet sustainable development objectives by overcoming exploitation in tourism. Besides, the International Social Tourism Organization (ISTO; https://isto.international/), established in 1963 as the International Bureau of Social Tourism (BITS), is an international non-profit organization whose mission does focus on social tourism. It includes many activities, as well as accessible, responsible, solidary, fair, and community-based tourism (Diekmann & McCabe, 2011). This organization and the purposes of social tourism have been recognized by the UNWTO.

The Montreal Declaration of 1996, *Towards a Humanist, Social Vision of Tourism*, and its addendum (2006), *Towards a Tourism Based on Development and Solidarity*, are ISTO landmark texts whose spirit is strongly connected with the principles of humanistic management in a broad perspective (Melé, 2016), as well as with the area of action of humanistic tourism.

Chapter 1 of this declaration states that social tourism aims to overcome discrimination and inequalities related to tourism as weel as the challenges of integration. This idea is consistent with some tourism studies which consider social tourism as a tool to reduce inequalities (Higgins-Desbiolles, 2006) and recognize human dignity as an enabler of morality under the circumstances of injustice or unfairness (Waldron, 2013). Accordingly, the declaration states that "all tourism development initiatives should be aimed at the full realization of each individual's potential, both as a person and as a citizen" (Art. 2), meaning a visitor and a host community member. Personal fulfillment is achieved through the development of human interactions – the discovery of new environments, cultures, and civilizations – and through the development of responsibilities taken on freely (Art. 4). In so doing, social tourism is an effective "shaper of society" that promotes social cohesion while driving economic growth (Chapter 2) and fostering sustainable development (Art. 8). In strict connection with justice, equality, and fairness, the whole community must benefit from economic development due to tourism through access to jobs and decent work (Art. 6). All tourism stakeholders are subject to the same expectations regarding their skills, capacities, and performance to ensure that their operations create more value for, and further benefit, host populations (Art. 7). Concerning local economic, social, and cultural benefits, tourism "reconciles tourism development, environmental protection, and a respect for the identity of local communities; bring[s] fresh resources into neglected regions; [and] promote[s] development without depletion of resources" without undermining the identity and value of local communities (Art. 8).

Due to the relevance of social tourism as a "shaper of society", a "driver of economic growth", a "driver of sustainable regional planning and local development", and a "partner in global development programs", other international organizations have also turned their attention to social tourism. The tourism policy of the European Union considers social tourism as an enabler of social integration, sustainability, and economic regeneration, representing the ideals, aims, and objectives of a truly social Europe (Diekmann & McCabe, 2011). Diekmann and McCabe (2011) also noted that, from the 2008 European Union Conference, based on the Lisbon Strategy, there is "a significant shift in orientation towards economic position on social tourism which is far removed from the earlier focus on social tourism as workers' rights and for individual and social benefits" (p. 422). In other words, social tourism may be considered as a humanistic form for managing societal well-being (Pirson, Bachani, & Blomme, 2018).

Through the focus on social tourism, these international organizations clearly mirror the interlinked micro, meso, and macro levels of interest of humanistic tourism, which are consistent with an information-intensive industry and an "experience good" resulting from the relationship and co-creation processes involving hosts and guests in different political, socio-economic, cultural, ethnic, and environmental contexts. These contexts play a role in how dignity, values, and norms are experienced in tourism, affecting the development and professionalization of humanistic tourism research, practice, policy, and pedagogy.

The challenges of humanistic tourism are manifold, both in terms of overcoming discrimination and inequalities in tourism and fostering prosperity and well-being. Decent work is an important issue, as tourism is a labor-intensive industry where employment opportunities for informal workers, the less skilled, minorities, women, migrants, and young people are higher than in most other industry sectors. Dignity violations are more likely in settings of unequal power, including financial, educational, social, cultural, hierarchical, and emotional disparities (Jacobson, 2009).

Strictly connected are the issues of justice, fairness, equity, and ethics in tourism development: They include the meaningful participation of minorities, indigenous people, and local communities in decision-making processes related to tourism planning and management; community engagement in the definition and communication of the destination tourism product when it involves local natural and cultural resources; and the distribution of the outcomes of tourism development. Under-participation, marginalization, and exclusion have effects on economic opportunities, culture, and relationships with the natural environment, affecting community well-being and dignity (Camargo & Vázquez-Maguirre, 2020).

Transversal challenges that remain largely unaddressed are climate change and biodiversity loss, which in turn are connected to pandemic crises, such as the current COVID-19 pandemic (Martin, 2006). Tourism growth places sharp environmental and social pressures on fragile natural ecosystems through the growing mobility and the concentration of people in some places; the development of global transport networks, as well as production and consumption processes (including those related to industrialized food production in global value chains); the consumption of resources; and the generation of waste, pollution, and CO_2 emissions. Some of the processes affecting climate change have also been recognized as the main reasons for the increasing pandemic threat in the 21st century. “Disease outbreaks such as SARS, Ebola, Marburg, hantavirus, Zika, and avian influenza are all outcomes of anthropogenic impacts” (Gössling et al., 2020, p. 7). In particular, “high-risk areas for the emergence and spread of infectious disease are where … wild disease reservoirs, agricultural practices, and cultural practices that increase contact between humans, wildlife, and livestock [occur]”. The pandemic has focused people’s attention on the connection between healthy ecosystems and human health and well-being, and it has highlighted how human beings are reliant on nature for their self-fulfillment, well-being, and sustainable development.

In conclusion, facing the economic challenges of contemporary society requires a humanistic culture focusing on human dignity and well-being, ethics, and social justice (Pirson & Turnbull, 2011). Tourism and destination management constitute interesting domains in which to experiment with humanistic management and address the challenges that threaten the achievement of sustainable tourism development (Higgins-Desbiolles et al., 2019).

6 Concluding Remarks

Previously existing and new challenges urgently require a move to a kind of post-capitalist world that fosters economic, social, and environmental sustainability, as well as the flourishing of human beings, society, and ecosystems (Higgins-Desbiolles, 2018; Pololikashvili, 2018). Humanistic tourism aims to address and foster this transition in the economy and society (Dettori & Floris, 2019; Giudici, 2018) by recognizing and embedding human dignity (Pirson, 2019) in tourism and destination management models.

Expanding on Cassano's meridian thinking (1997), tourism is at the "frontier", an area in which creativity and knowledge thrive and drive transformation. "The frontier is where consolidated identity comes into contact with what lies outside. Therefore, the periphery [is] where everything ends and where the waste of the center rolls, but [it is] an epistemologically higher place" (Cassano, 1997, p. 28). Meridian thinking is an inspiring way to move away from established models or paradigms of tourism development and to move toward an emerging one, thus allowing for a "continuous start over". According to the meridian thinking, a society, which has profit at its core, is to be considered unwise.

References

Ajagunna, I. (2014). New world, new tourism, new destinations. Implications for emerging economies and lessons for "old" tourism-dependent developing economies. *Worldwide Hospitality and Tourism Themes*, 6(2), 103–117. doi:10.1108/WHATT-12-2013-0049.

Andriotis, K. (2018). *Degrowth in tourism: Conceptual, theoretical and philosophical issues*. Oxfordshire, UK: CABI Publishing.

Bocchi, G., & Ceruti, M. (Eds.). (2007). *La sfida della complessità*, *65*. Italia Spa: Pearson.

Bramwell, B., & Lane, B. (2011). Critical research on the governance of tourism and sustainability. *Journal of Sustainable Tourism*, *19*(4–5), 411–421. doi:10.1080/09669582.2011.580586.

Burnes, B. (2005). Complexity theories and organizational change. *International Journal of Management Reviews*, 7(2), 73–90. doi:10.1111/j.1468-2370.2005.00107.x.

Butler, R. (2015). The evolution of tourism and tourism research. *Tourism Recreation Research*, *40*(1), 16–27. doi:10.1080/02508281.2015.1007632.

Camargo, B. A., & Vázquez-Maguirre, M. (2020). Humanism, dignity and indigenous justice: The Mayan train megaproject, Mexico. *Journal of Sustainable Tourism*, 1–20. doi:10.1080/09669582.2020.1758707.

Cassano, F. (1997). Sapere di confine. La frontiera come luogo epistemologicamente più alto. *Pluriverso*, *1*, 22–30.

Chandler, A., Jr. (1963). *Strategy and structure. Chapters in the history of American industrial enterprise*. Cambridge, MA: MIT Press.

Clarke, J. (1997). A framework of approaches to sustainable tourism. *Journal of Sustainable Tourism*, *5*(3), 224–233. doi:10.1080/09669589708667287.

Della Lucia, M. (2018). *Approccio manageriale allo sviluppo turistico sostenibile. Un framework interpretativo e di gestione integrato*. Milano: Franco Angeli.

Della Lucia, M., & Franch, M. (2017). The effects of local context on World Heritage Site management: The dolomites natural world heritage site, Italy. *Journal of Sustainable Tourism*, *25*(12), 1756–1775. doi:10.1080/09669582.2017.1316727.

Della Lucia, M., & Martini, U. (2012). Destination governance. Marginal rural areas in Europe. In E. Fayos-Sola, J. Jafari, & J. A. M. de Silva (Eds.), *Knowledge management in tourism: Policy and governance applications* (pp. 265–281). Bingley, UK: Emerald Group Publishing Limited.

Dettori, A., Caboni, F., & Giudici, E. (2017). Education and sustainability: The case of emotions park. *Journal of Management and Sustainability*, *7*(4), 1–11. doi:10.5539/jms.v7n4p65.

Dettori, A., & Floris, M. (2019). Sustainability, well being, and happiness: A co-word analysis. *International Journal of Business and Social Science*, *10*(10), 29–38. doi:10.30845/ijbss.v10n10a5.

Diekmann, A., & McCabe, S. (2011). Systems of social tourism in the European Union: A critical review. *Current Issues in Tourism*, *14*(5), 417–430. doi:10.1080/13683500.2011.568052.

Dodds, R., & Butler, R. W. (Eds.). (2019). *Overtourism: Issues, realities and solutions*. Berlin, Boston: De Gruyter.

Fayol, H. (1917). *Administration industrielle et generale*. Paris: Dunod et Pinat.

Franch, M., Martini, U., Buffa, F., & Parisi, G. (2008). 4L tourism (landscape, leisure, learning and limit): Responding to new motivations and expectations of tourists to improve the competitiveness of alpine destinations in a sustainable way. *Tourism Review*, *63*(1), 4–14. doi:10.1108/16605370810861008.

Giudici, E. (2018). Quali paradigmi per l'impresa del terzo millennio? In A. A. *Scritti seri e semiseri in onore di Claudio Baccarani* (pp. 201–206). Torino: Giappichelli.

Giudici, E., Dettori, A., & Melis, C. (2012). Parks – A perfect model of sustainable tourism? *The 2nd International Conference on Tourism Management and Tourism Related Issues*. Nice, France.

Godfrey, K., & Clarke, J. (2000). *The tourism development handbook: A practical approach to planning and marketing*. London: Cassell.

González Tirados, R. M. (2011). Half a century of mass tourism: Evolution and expectations. *The Service Industries Journal*, *31*(10), 1589–1601. doi:10.1080/02642069.2010.485639.

Gössling, S., Scott, D., & Hall, C. M. (2020). Pandemics, tourism and global change: A rapid assessment of COVID-19. *Journal of Sustainable Tourism*, 1–20. doi:10.1080/09669582.2020.1758708.

Hall, C. M. (2008). *Tourism planning. Policies, processes and relationships*. Harlow, UK: Pearson.

Hall, C. M. (2009). Degrowing tourism: Decroissance, sustainable consumption and steady-state tourism. *Anatolia*, *20*(1), 46–61. doi:10.1080/13032917.2009.10518894.

Hall, C. M. (2010). Changing paradigms and global change: From sustainable to steady-state tourism. *Tourism Recreation Research*, *35*(2), 131–143. doi:10.1080/02508281.2010.11081629.

Hamel, G. (2007). *The future of management*. Boston: HBS Publishing.

Higgins-Desbiolles, F. (2006). More than an "industry": The forgotten power of tourism as a social force. *Tourism Management*, *27*(6), 1192–1208. doi: 10.1016/j.tourman.2005.05.020.

Higgins-Desbiolles, F. (2018). Sustainable tourism: Sustaining tourism or something else? *Tourism Management Perspectives*, *25*, 157–160. doi:10.1016/j.tmp.2017.11.017.

Higgins-Desbiolles, F. (2020). Socialising tourism for social and ecological justice after COVID-19. *Tourism Geographies*, 1–14. doi:10.1080/14616688.2020.1757748.

Higgins-Desbiolles, F., Carnicelli, S., Krolikowski, C., Wijesinghe, G., & Boluk, K. (2019). Degrowing tourism: Rethinking tourism. *Journal of Sustainable Tourism*, *27*(12), 1926–1944. doi:10.1080/09669582.2019.1601732.

Hitt, M. A. (2000). The new frontier: Transformation of management for the new millennium. *Organizational Dynamics*, *28*(3), 7–17. doi:10.1016/S0090-2616(00)88446-6.

Hockings, M., Dudley, N., Elliott, W., Napolitano Ferreira, M., MacKinnon, K., Pasha, M. K. S. & Yang, A. (2020). Editorial essay: COVID-19 and protected and conserved areas. *PARKS*, *26*(1), 7–24.

Inskeep, E. (1991). *Tourism planning*. New York: Van Nostrand Reinhold.

Ivan, O. (2017) 'We make more money now, but we don't talk to each other anymore': On new tourism and capitalism in the Danube Delta. *Journal of Tourism and Cultural Change*, *15*(2), 122–135. doi:10.1080/14766825.2016.1260102.

Jacobson, N. (2009). A taxonomy of dignity: A grounded theory study. *BMC International Health and Human Rights*, *9*(1), 3. doi:10.1186/1472-698X-9-3

Kallis, G. (2011). In defence of degrowth. *Ecological Economics*, *70*(5), 873–880. doi:10.1016/j.ecolecon.2010.12.007.

Kallis, G., Kostakis, V., Lange, S., Muraca, B., Paulson, S., & Schmelzer, M. (2018). Research on degrowth. *Annual Review of Environment and Resources*, *43*(1), 291–226. doi:10.1146/annurev-environ-102017-025941.

Latouche, S. (2014). Imaginary, decolonization of. In G. D'Alisa, F. Demaria, & G. Kallis (Eds.), *Degrowth: A vocabulary for a new era* (pp. 117–120). New York: Routledge.

Lee, Y. K., Lee, C. K., Choi, J., Yoon, S. M., & Hart, R. J. (2014). Tourism's role in urban regeneration: Examining the impact of environmental cues on emotion, satisfaction, loyalty, and support for Seoul's revitalized Cheonggyecheon stream district. *Journal of Sustainable Tourism*, *22*(5), 726–749. doi:10.1080/09669582.2013.871018.

Lew, A. A. (2011). Tourism's role in the global economy. *Tourism Geographies*, *13*(1), 148–151. doi:10.1080/14616688.2010.531046.

Martin, J. (2006). *The meaning of the 21st Century*. New York: Penguin Group.

Mayo, E. (1933). *The human problems of an industrial civilization*. Oxford: Routledge.

Melé, D. (2003). The challenge of humanistic management. *Journal of Business Ethics*, *44*(1), 77–88. doi:10.1023/A:1023298710412.

Melé, D. (2012). The firm as a "community of persons": A pillar of humanistic management ethos. *Journal of Business Ethics*, *106*(1), 89–101. doi:10.1007/s10551-011-1051-2.

Melé, D. (2016). Understanding humanistic management. *Humanistic Management Journal*, *1*(1), 33–55. doi:10.1007/s41463-016-0011-5.

Milano, C., Novelli, M., & Cheer, J. M. (2019). Overtourism and degrowth: A social movements perspective. *Journal of Sustainable Tourism*, *27*(12), 1857–1875. doi:10.1080/09669582.2019.1650054.

Novelli, M. (2015). *Tourism and development in Sub-Saharan Africa: Current issues and local realities*. London: Routledge.

Parsons, H., Houge Mackenzie, S., & Filep, S. (2019). Facilitating self-development: How tour guides broker spiritual tourist experiences. *Tourism Recreation Research*, *44*(2), 141–152. doi:10.1080/02508281.2019.1582159.

Pedersen, S. B. (2020). A passport to peace? Modern tourism and internationalist idealism. *European Review*, 1–14. doi:101017/s1062798719000516. Retrieved from http://www.cambridge.org/core.

Pirson, M. A. (2019). A humanistic perspective for management theory: Protecting dignity and promoting well-being. *Journal of Business Ethics*, *159*(1), 1–19. doi:10.1007/s10551-017-3755-4.

Pirson, M. A., Bachani, J., & Blomme, R. J. (2018). *Humanistic management: Social entrepreneurship and mindfulness*, Vol. II. New York: Business Expert Press.

Pirson, M. A., & Lawrence, P. R. (2009). Humanism in business – Toward a paradigm shift? *Journal of Business Ethics*, *93*(4), 553–565. doi:10.1007/s10551-009-0239-1.

Pirson, M., Martin, K., & Parmar, B. (2017). A humanistic perspective for management theory: Protecting dignity and promoting well-being. *Journal of Business Ethics*, *159*(1), 1–19. doi:10.1007/s10551-017-3755-4.

Pirson, M. A., & Turnbull, S. (2011). Toward a more humanistic governance model: Network governance structures. *Journal of Business Ethics*, *99*(1), 101–114. doi:10.1007/s10551-011-0752-x.

Pololikashvili, Z. (2018). The road ahead: Together for a more competitive and responsible tourism sector (new year message). Retrieved from http://www2.unwto.org/content/road-ahead-together-more-competitive-and-responsible-tourism-sector-new-year-message.

Pung, J. M., Gnoth, J., & Del Chiappa, G. (2020). Tourist transformation: Towards a conceptual model. *Annals of Tourism Research*, *81*, 1–12. doi:10.1016/j.annals.2020.102885.

Russo, A. P., & Richards, G. (Eds.) (2016). *Reinventing the local in tourism*. Clevedon, UK: Channel View.

Schneider, F., Martinez-Alier, J., & Kallis, G. (2011). Sustainable degrowth. *Journal of Industrial Ecology*, *15*(5), 654–656. doi:10.1111/j.1530-9290.2011.00388.x.

Scott, R. W. (2003). *Organizations. Rational, natural, and open systems*. Thousand Oaks, CA: SAGE.

Secchi, D. (2011). *Extendable rationality. Understanding decision making in organizations*. New York: Springer.

Secchi, D., & Newmann, M. (Eds.) (2015). *Agent-based simulation of organizational behavior. New frontiers of social science research*. New York: Springer.

Smith, M. K., & Diekmann, A. (2017). Tourism and wellbeing. *Annals of Tourism Research*, *66*, 1–13. doi:10.1016/j.annals.2017.05.006.

Spitzeck, H. (2011). An integrated model of humanistic management. *Journal of Business Ethics*, *99*(1), 51–62. doi:10.1007/s10551-011-0748-6.

Swarbrooke, J. (1999). *Sustainable tourism management*. Wallingford, UK: CABI Publishing.

Taylor, F. W. (1911). *The principles of scientific management*. New York: Norton.

Waldron, J. (2013). The paradoxes of dignity – About Michael Rosen, dignity: Its history and meaning (Harvard University Press, 2012). *European Journal of Sociology*, *54*(3), 554–561. doi:10.1017/S0003975613000404.

Winchenbach, A., Hanna, P., & Miller, G. (2019). Rethinking decent work: The value of dignity in tourism employment. *Journal of Sustainable Tourism*, *27*(7), 1026–1043. doi:10.1080/09669582.2019.1566346.

Part I

Criticism and Ethical Value in Tourism

2 The Truths about Conventional Tourism and Its Scientism

A Humanistic Critique[1]

Atila Yüksel

1 Introduction

Conventional tourism (CT), described as "*convenience, undifferentiated marketing, mass-consumed experiences centered on pleasure principle*" (Kontogeorgopoulos, 2004, p. 87), is expected to bring "peace and prosperity". However, this has not happened so far, as there is more than initially meets the eye. Many other new and more sustainable, ecological forms of tourism have been devised and launched. But, as Butler (1999, p. 12) notes "*it is almost impossible to have a form of tourism development that does not impact on the location in which it occurs*". CT and other forms of tourism are not mutually exclusive at all (Butler, 1999). CT strongly connects with such forms of "mass ecotourism" and "sustainable tourism". At present, CT is still popular and it continues to manifest itself as a strong construct of western hegemony, containing western ideologies, interests, values, and practices (Cater, 1995). CT closely resembles imperial colonial tools (Palmer, 1994), used in the process of *touris(m)tification* by substituting a place's social, cultural, natural, psychological, and commercial identity, and fabric for the exploitation of the tourists and capitalization of the tourism industry. And it holds the potential to turn into a burden even for the developed world (Jover & Díaz-Parra, 2019; Minquez, Pineria, & Tabales, 2019; Sequera & Nofre, 2018). CT follows the laws of capitalism in order to capitalize on the demand for *hedonism, leisure, and entertainment* for making profits. It is a way of colonizing and commodifying the life, culture, heritage, environment, etc., of local people for the gain of both economic and ideological profits (Palmer, 1994). CT values the "independence" of the instrumentalist tourist who wants to feel like a king or queen for a few days. It is mostly built on inequality and on wealthy and bourgeois tourists' exploitation of the way of, often primitive, the life of locals, their cultural customs, heritage, etc. In contrast to its most promoted goals, such kind of so-called interchanges as cultural, educational, and/or spiritual rarely transpire in the CT context. And interestingly, conventional research often neglects this dehumanizing and exploitative nature of tourism, though unintentionally, by focusing on attributes and issues that enhance organizational outputs. While CT fosters curiosity about other cultures and their primitive life for self-gratification, it does not pursue the goal of making the local's world more just. Is there any dignity in this?

Considering the growing annual volume of international tourist arrivals, the generated total revenue, and the jobs, it is not surprising to witness that governments in developing countries take CT as a panacea to remedy economic problems (Benavides, 2001; Burns & Novelli, 2008). But this is *delusional,* because CT is not harmless. And it is not as "innocent" as it has been long claimed. Regardless of the noun or adjective that precedes the word of "tourism", all tourism types are likely to consume places, create waste, require infrastructural development, compete for scarce resources, over-exploit resources, prioritize profits and disguises costs, stimulate irresponsible behaviors, degrade the environment, and segregate locals (Butler, 1999; Sharpley, 2009). The current conceptualization of CT as "traveling and staying" is somewhat *hypocritical* as it disguises likely inhumane realities (*touristification, gentrification, marginalization, commercialization, etc.,*) implicit in this mass movement. This distracts us from dealing with the real problems of *touris(m)tification* through CT by focusing attention on wrong issues (Robinson, 2004). The problem with the CT does not lie solely in its mismanagement or in excessive exploitation of local resources. It lies in modern societies encouraging people to travel in greater numbers and more frequently (Sharpley, 2009). As Krippendorf (1987) suggested long ago, reducing the need to escape periodically from the modern society is possible through improving people's day-to-day social well-being. Hence, despite its so-called promoted benefits, *touris(m)tification* in the form of CT may not be the best alternative to reduce poverty, especially in the less-developed countries (Bigsten & Shimeles, 2007; Buhalis, 1999). Moreover, in line with Botterill (1989), I suspect that there exists a kind of a "self-indulgence" in CT research. Traditionally speaking, positivism is seen as the only beacon for "*predicting*" cause-and-effects in CT, but tourism is in need of more understanding than predictions.

This chapter is devised to draw attention to the issue that while many destinations rely heavily on conventional forms of tourism for the generation of income and employment, CT and its scientism can be a source of problems as well, particularly when they are not well planned, managed, and developed. As with any "*ism*" which generates its own course of and "*endism*", CT and its scientism have produced a set of imperialistic beliefs, attitudes, and measurement approaches to explain how tourism and its research "ought" to be. *Scientism* refers to the matter of putting too much value only on science in comparison with other branches of learning (Sorell, 1991). Not only the costs and problems associated with CT (i.e., volume) and but also its current scientism warrant an extensive debate. Drawing on Botterill's (1989) criticism on "who benefits from tourism and tourism research", this chapter first outlines delusions, misconceptions, and hypocrisies in the business of CT. Then, humanistic tourism (HT) is introduced and likely implications of vacations on tourists are addressed. The next section deals with whether CT and its research should just be for creating and supporting an industry, for the accumulation of information enabling structural change, or for the well-being of tourists, natives, and researchers who are in search of "self" (Botterill, 1989). Finally, a call for humanistic research is made to extend beneficial outcomes for all parties involved in the tourism community.

2 Volume of the CT: Pros and Cons

Several adverse consequences of haphazard tourism enlargement have been reported, including environmental, economic, and socio-cultural (Tosun, 2000). More specifically, the CT and its generally overlooked negative consequences mostly focused on locals. These include limited exchanges between hosts and the visitors akin to master-slave relationships, rising visitor numbers and diminishing carrying capacities of destinations, as well as, environmental degradations, social segregations resulting from overtourism, gentrification, etc., (Jover and Díaz-Parra, 2019; Minquez et al., 2019; Sequera & Nofre, 2018). Previous studies suggest that tourism destinations, particularly in developing countries are not immune to pressures brought by the CT (Benavides, 2001; Burns & Novelli, 2008; Hundt, 1996; Jover and Díaz-Parra, 2019; Minquez et al., 2019; Sequera & Nofre, 2018; Yuksel, 2019a). Therefore, they are likely to face many known, as well as, unprecedented problems. Considering some of the developmental issues well-evidenced in the process *touris(m)tification*, the chance to experience one of the problems listed below by destinations in the developing countries is highly likely:

i Foreign capital takes over domestic capital. This results in a change in the status of ownership and hence the attractiveness of locality (e.g., naturalness, localness, heritage, etc.) is expended for the sake of urbanization and rapid economic returns. *Touris(m)tification* through foreign capital taking over local capital is likely to bring in growth, "economic enlargement" without development (Ghatak 1995);
ii Economic enlargement often involves construction and alterations for pleasing outsiders as opposed to all-year-round residents. And locals are generally employed in low-skill positions. Under these circumstances, the possibility of skill-upgrading through *tourismtification* is unlikely to happen. As locals remain underqualified, foreign operators will push for a migrant labor force for achieving higher service quality. And this migration will eventually generate resentment due to social injustice and

income leakages. Employment in the tourism industry will often be dehumanizing as it demands displays of "friendliness and servitude" that locals may find it insulting (Fouberg, Murphy & Blij, 2015);

iii Dehumanizing nature of *tourismtification*, with a very narrow economic perspective by commodifying local cultural and natural assets, will result in congestion, overcrowding and conflicts in "open social museums" in which wealthy and bourgeois visitors enjoy their time while staying oblivious to the real problems of those who are kept in the social prison against their free will (Yüksel, 2019a).

At the first glance, the aforementioned problems may be taken peculiar only to the developing world; however, evidence from different developed countries pertaining to resistance to tourism shows that this may not be the case. Voices continue to rise from France, Italy, Spain, Holland, echoing that reckless tourism, whether conventional or not, holds the potential to make life miserable, particularly for residents and locals as employees (Egresi, 2018; Fouberg, Murphy & Blij, 2015; Jover & Díaz-Parra, 2019; Minquez et al., 2019; Sequera & Nofre, 2018; Zanardi, 2019). In the midst of current uncertainty resulting from Covid-19 pandemic, such statements as "*Tourism kills the city, we are sick of gentrification*" (Egresi, 2018) and "*Tourist: your luxury trip is my daily misery*" (Zanardi, 2019), underline the fact that humanistic tourism and what it really triggers in physical, mental, and spiritual states of individuals will become the main subject of future research.

3 Humanistic Tourism (HT)

Touris(m)tification, driven by profit greediness, has reached a point where tourism has become related to "objects only", not to the individuals or to their life. There are some forms of CT in which no income reaches local hands, and this has forced the birth of new forms of respectful social enterprises (e.g., Planetterra, Intrepid Foundation) to address the problems of communities who have never been a part of the tourism industry. Despite these genuine efforts, anecdotal and research evidence suggests that tourists at large, though they have comparatively enough income to support local economies and families, often treat locals as objects like in a museum but not as subjects. Instead of interacting with natives, the general tendency is to keep a safe distance for taking a general picture. And this suggests that some tourists are likely to keep a social distance and stay as an observer behind the restricted camera lens in order not to be disturbed by what is really going on locally (Errington & Gewertz, 1989; O'Rouke, 1987). Several dehumanizing examples were reported from the developing world: for example, some males from the western world, aiming to have an opportunity of "good time", continue to flock into Asian destinations to "buy" under-aged Asian girls, who are often unable to communicate verbally, for sexual encounters, and all human-related attributes such as names, values, family history are not exchanged in these encounters (Pritchard, 2014). The lives of locals are ruined because all male tourists, without any exception, want to feel like a (hu)man. Hence it would not be wrong to claim that *touris(m) tification,* in most of the cases, deprives people of positive human qualities, as a broad range of indignities exist. Subtle (looks, gestures, tone of voice, disrespect to local customs, patronizing interactions, etc.) or brutal forms of dehumanization occur not only when one group of people (tourists) see others as objects and not as human, but also when the powerful (tourists) take the very humaneness of others by trying to manage their economic and social relations. Implying that CT and dehumanization are the two sides of the same coin should not be an overstatement.

As a result, a new concept – humanistic tourism (HT) – has been coined (Botterill, 1989). It gently reminds us of the simple fact - we must return to being a "human", aiming to live a meaningful life. Humanism in tourism is much needed as humanism indeed can defend human dignity in the face of its vulnerability. To defend human dignity means to respect human freedom and be against the instrumentalization of human beings into human capital and mere means for profits. In this respect, HT seems to hold potential as an experiential tourism activity. However, can HT challenge the present dominant neoliberal model? As it is, the HT flourishes under the shadow of CT. Thus, it includes both characteristics of CT and its own unique characteristics, particularly the moral aspects with an aim to help people fulfill their potential and maximize their persona development and well-being (Caton, 2012). Some view HT as "*a revolutionary way of reducing the negative impacts of tourism and takes into account the relationship between hosts and guests as based on the tourist experience*" (Bizzari, 2016, p. 110.) To some, HT concerns with the moral world of tourists to achieve sustainable tourism, and people's life, historical relics, culture and arts, ethnic customs, cultural landscapes, social environment, etc., are therefore can be taken as HT resources (Zhou, 2017). Similarly, La Lucia and Giudici (2020) emphasize the reciprocal respect between the host and the visitor and note that in HT "*tourist activities must assume a configuration capable of respecting not only the environment but also the culture of the places with the involvement of the residents*". HT is expected to promote restoration of the human person via solidarity, responsibility, and social unity in order to contribute to dialogue and mutual knowledge (Bizzari, 2016; Goytia & de la Rica, 2014). The CT is characterized by pursuing pleasure and avoiding pain for self-gratification, in contrast HT seems to support the idea of "delayed gratification" for self-actualization in a socially responsible way. However, before moving into this discussion, we must first make sure that *whether conventional tourism activities are any beneficial to the individual at all.*

3.1 Vacation: A Trauma or Recovery?

On one hand, we have several studies showing that deteriorated well-being stemming from persisting work stress can be fixed through a vacation (Fritz & Sonnentag, 2006). Taking a break from work, both mentally and physically, is argued to be helpful in maintaining high performance at work. According to Sonnentag (2003), psychological detachment from work during off-hours is beneficial for relaxation, life satisfaction, and absence of burnout. Similarly, Fritz & Sonnentag (2006) note that having built up resources during a vacation may allow the employee to work more efficiently, thereby increasing daily task performance. In another study Fritz & Sonnentag (2005) found that weekend experiences could predict different aspects of job performance after the weekend. Similarly, Sonnentag (2003) reported that individuals who felt that they sufficiently recovered during leisure time experienced a higher level of work engagement during the subsequent workday. Since temporary breaks from work can influence the productivity, performance, health, and/or life-satisfaction of the individual, the concept of vacation has become a topic of interest in health and management research (Strauss-Blasche, Ekmekcioglu, & Marktl, 2000). The "paid-vacation" has hence become a legal right in many countries, except the USA.

On the other hand, there are several other studies implying that vacation can be a trauma (see Eden, 1990; Eden, 2001; ENP, 2014; Gump & Matthews, 2000; Hartig, Catalano, Ong, & Syme 2013; Kop, Vingerhoets, Kruithof, & Gottdiener, 2003; Tarumi & Hagihara, 1999; Tonelo, Providência, & Gonçalves 2013; Van Heck & Vingerhoets 2007; Yüksel 2016). The results of De Bloom et al.'s studies (2009, 2010a, 2010b,) suggest that in contrast to what has been claimed previously vacation has only small positive effects on health and well-being. And that these effects soon fade out after work resumption. Mood and quality of sleep lasted only up to five weeks (Strauss-Blasche, Ekmekcioglu, & Markti, 2000), burnout from work returned to pre-vacation levels after three to four weeks (Westman & Eden 1997; Westman & Etzion 2001), the impact of a holiday on post-trip life satisfaction is small (De Bloom et al. 2009), positive effects on well-being and performance-related outcomes return to their pre-vacation level three or four weeks after vacation (Eden, 1990; 2001). The positive effects of vacation diminish within a short while after the employee's return to work (Fritz & Sonnentag, 2006). Hence no matter how much a vacation has been enjoyed, there exists a "fade out" period – where all the effects of the vacation dwindle rapidly.

Vacations can indeed be a stressful period as it breaks ordinary biorhythms in general. Emotional, physical, and mental body cycles are to be affected by daily regularities. And vacations are likely to disrupt these cycles. A cycle lasts about 23 to 33 days. A new cycle (or breaking an already existing one) is started whilst on holiday to be disrupted after returning home (or vice versa). Vacation is a highly susceptible commodity that can alter almost everything in bio-chemistry by slowing or rushing things. Some researchers even suggest that the possible negative effect of vacation on human health begins at least one week before taking a vacation. Those first days of vacation are associated with migraine, nausea, muscle spasms of unknown origin, and viral infections leading to fever, cough, cold, and influenza. 3% of the Dutch who have taken vacations suffer from these disorders mildly or seriously in the first days of vacation. The effect of going to another place for vacation, that is "(the anxiety of) taking vacation", results in an increase of the blood pressure one day before the vacation. The high pressure continues on the day of travel and on the first day of the vacation. The insomnia in the first night is almost inevitable. Only after the fifth day of vacation, the body restores itself. Vacations also cause shifts in mood and the state of mind (Nawijn, 2011a, 2011b; Nawijn, De Bloom, ve Geurts, 2013a; Nawijn, Marchand, Veenhoven, & ve Vingerhoets, 2010; Nawijn et al., 2013b; Yüksel, 2016). Tourists reported poorer moods and more health complaints during the first three days of their visit (Pearce, 1981). Travelling, temporally moving away from the familiar surroundings, neighborhood, and friends are all a source of stress on their own. There are physiological and psychological consequences of struggling against this stress. 4% of medical cases that happen in flights are connected with mental disorders related to vacation. 90% of them are related to anxiety (flight fear, panic attack, social phobia, etc.). The death rate in the first two–three days of vacation is higher when compared to the following days of vacation. This can be the result of the transmission from lively daily life (life and work) to unusual tranquility (e.g., calmness). Think of this way: one shifts from the highest gear to the lowest one or vice versa.

"The Post-Holiday Memory Loss Syndrome" (ENP, 2014) (especially towards knowledge or skills), which is common in children returning back to school after summer holidays, can probably be generic to those who come back to work after vacation. Vacations can even be a nightmare after returning for some due to forgotten passwords, login codes of e-mail accounts, bank accounts, and phone, unpaid bills, holiday credit debits, etc. This may result from temporary suspension of the correspondences with these passwords during the vacation period. A great effort that should be made in order to catch up on the backlog of workload that has accumulated is straining. Similarly, the time and effort needed to remember and re-catch up with the things that have been temporarily postponed along with daily work habits and relationships that have been abandoned during vacation (e.g., forgetting how to run the package programs, etc.) could turn out to be stressful. Other stressful occasions include psychological costs of those hallow gazes in the first meetings after a vacation, homesickness (Kop et al. 2003; Pearce 1981; Van Heck & Vingerhoets 2007; Vingerhoets, van Huijgevoort, & van Heck, 2002), likely feeling of guilt (!) that the workaholics feel during vacation and disharmony (homeostatis disorder) created by the body which does not need the hormones that help to deal with work stress it excretes while being on vacation.

Vacation is undoubtedly an idealized moment (Yüksel, 2016). It naturally includes great expectations and fantasies. And it includes not only monetary but also all types of nonmonetary psychological costs of frustrations caused by the failure to meet these expectations. It cannot be expected that the psychological distress and debates triggered by both the frustration due to the unfulfilled expectations and by the accusation of others (family members, relatives) or of oneself will have a positive effect on an individual's health. Since holiday is a fantasy, people may attempt to assume the identities that do not belong to them. It would be surprising to consider that the anxiety or sense of loss of a pet or a favorite object would not be displayed toward the "*identity lost*" during a vacation. Vacations can become tiresome and exhausting for families with children. It is not so difficult to understand that vacation is a stressful moment and also to figure out why the positive effects of vacation disappear so quickly considering the couples that quarrel more in vacation (Ryan 1991) due to issues of changing identity and character.

Vacation can be a tiring period for double transformation. There are not only clothes in people's luggage on a journey. There are psychological and psychical habits that they carry everywhere at the same time. Many vacationers have to take a vacation in the areas and regions which do not suit their own identities and habits. It is certain that the stress felt by a person who gives up doing sports during vacation will affect his/her health negatively as s/he has the habit of going to the gym in his/her routine every day. Certainly, the relationship between vacation and an individual's well-being and productivity can differ according to the tourism activity, working customs (such as advanced technology), free time for one's self, time to travel to the destination, interpersonal conflicts, climate conditions (warmer and sunnier), vacation locations, exercise and activities during a vacation (resource consuming versus resource building activities), good sleep, duration and frequency of vacation, health conditions of the individual and making new acquaintances of individuals, etc. Further research is needed to deal with the question of whether "humanistic tourism, designed to meet personal growth needs, better influence vacationers' well-being and life satisfaction than the traditional activities of the CT?"

3.2 Humanism in Tourism Research: Is It Another "Ism"?

Several tourist and tourism types have been formulated, from mass to independent, from eco-tourist to cultural tourist, gastronomic tourist, etc. While a new type of tourism, called "humanistic" is added to the list, there exists an interesting definitional discord as to what it is. The lack of definitional consensus of "who really is the humanistic tourism and tourist?" inevitably raises serious concerns about the direction of its research. *Is humanistic tourist a person or a product? Can humanistic tourists better socialize? Do they have more free will and the consciousness of a close relationship with another person? Do they have more intentions for social interactions and psychological reactions to inhumane events, incidents, etc., than other types of tourists?*

Traditionally, the tourist is defined according to "from-to" form of movement behavior (i.e., going from one place to another place and staying overnight). And the definition of humanistic tourist does not seem to be an exception in this respect. If there is a movement for visiting "beautiful" places, then what is the underlying distinction of being a humanistic tourist? Can "stationary" or "limited" movement be helpful in defining a humanistic tourist? That is, instead of overnight accommodation, requiring movement, humanistics, who are conscious about the effects of over-tourism on fragile lands and lives, may be motivated to prefer *staycation.* They may like to spend time at their already urbanized homeland, paying visits to local attractions without overnight staying elsewhere. In addition to the movement problem at hand, there is another issue. That is, if humanistics are to be described as those "*respecting human and taking humanistic activities on board whilst on vacation*" then we face a boundary issue in that "*what is humanistic for tourists may not be so humanistic for the host*".

In addition to the definitional issues, the rampant positivist research practices may constitute a big hurdle against better understanding humanistic tourism. The positivist research is far from putting "humanism" in the center. Instead, positivist research exploits the conditions, traumas, and psychology of the others for its own benefit. In this respect, positivist research is very conducive to creating fiction altruism that best works for organizations, not for locals. "Can imperial tradition of positivism be humanized?" This question is vital since humanistic research involves giving prime place to human beings, human meanings, and human actions in research (Lewis-Beck, Bryman, & Liao, 2004). Its two tenets include "(1) *all human knowledge ultimately represents interpretations of human experience; therefore it is important to take experience seriously, and to try to understand how the processes of interpretation function, and* (2) *an appreciation of the individual's unique experience is indispensable for a complete understanding of that person, particularly in clinical and growth facilitating relationships*" (Wamoth, 1998, p. 314).

I believe that the road before the HT research is long and thorny. The following are some of the challenges that researchers in HT may face due to cumbersome and rampant practices inherited from the positivist research. (1) Generally speaking, the positivist research hypothesizes in advance about the strength and the direction of the relationship among subjectively chosen dependent and independent variables. The researchers then engage participants accordingly with the predeveloped questions, experiments, etc. But, in general, the researchers do not share the results with participants and they never wonder what had happened to the subjects after the research was completed and published. Is this humane? Broadly speaking, the process of hypothesizing, data collecting, analyzing, interpreting, and finally publishing does not have any concrete benefit neither to the destination nor to the person who is under examination. When the research is over, researchers do not generally return to the people who were kindly forced, through research questions, to remember and talk about their current or past traumas they do not want to remember and they are then left with a deteriorated psychology. Almost no care exists as to whether the so-called fancy recommendations, produced as a result of the empirical research are applied to improve the conditions of the people examined. Thus, there exists a "selfishness, greediness, and exploitations" in positivist research in that the research is meant to gain status or personal prestige through publication. The purpose of positivist-empirical research is not to solve "real problems" of locals, say to increase the social welfare, and create a just world but to create a new "operable variable" for the business and the publishing industry. Should any obtrusive humanistic research be justified as a selfish act to get statues or personal credit through publication? Should the humanistic researcher have the right to distance himself from the subjects or the phenomenon as in the case of positivist research? How can the humanist researcher combine his personal liberty with social responsibility, personal fulfillment aspiring to the greater good of humanity? (2) The linear cause-and-effect tradition in the positivist approach suggests that even a social reality can be divided into its subdimensions, variables, causes, and effects. Positivism insists that there is only one absolute reality and this is often constructed by the researcher. Does this view fit into the humanistic approach? What happens if there are multiple realities constructed by the participants? (Hirschman, 1986) What if the reality continually develops at a differing logarithmic pace? And what if the reality cannot be divided? When faced with contradictory moral affairs, the conventional research paradigms, as they have long outlived their usefulness, are likely to fail to understand "the multiple human reality". As Buchanan (1998) notes, means and ends are not independent but rather they are inextricably intertwined. That is, how the researcher thinks about the research subject will give shape to how s/he will lead the research. If the choice of phenomenon, method, data, and findings is inherently value-laden, how anyone involved in humanistic research claim that the reality of the "researched" is captured objectively? (3) The tradition of generalizability in the CT research disregards the nonrecurring nature and uniqueness of the individual experience. Furthermore, business-driven research stressing that tourism has a generalizable positive effect on a destination subconsciously disables the society from the choice of and act on paths to freedom. So-called generalizable studies that talk about "tourism" as a paradise may prevent society from thinking and embarking on other possible economic endeavors. Equally importantly, how humanistic is it to reduce science to arguable statistics and claim that one has discovered an experience that could be generalizable across time and social context? (4) The "publish or peril" tradition dominant in the positivist tradition, along with misconducts in citations, may constitute further challenges for progress in humanistic research. Citing of publications positively which had already been retracted from journals long ago continues even still, at an accelerating rate (Halevi and Bar-Ilan, 2016). Alarmingly, almost one in five researchers does not read the original study that they are citing (Simkin & Roychowdhury, 2002). Can we understand and progress HT with this burdensome baggage? I suppose that leaving the traditions of CT behind and diving into a shining new world of humanistic tourism will not be that easy.

Conclusion

While many destination authorities rely heavily on tourism due to its likely effects on the generation of income and employment, CT and its "scientism" can be a source of problems as well, particularly when they are not well planned, managed, and developed. As with any "*ism*" which generates its own course of "*endism*", CT and its scientism have produced a set of imperial beliefs, attitudes, and measures to explain how tourism should develop and how its research "ought" to be. Certainly, no "*ism*" is free of its deficiencies and hence every "ism" comes to an end. In this chapter, an attempt was made to provide a humanistic critique of the inherent costs and problems of CT (i.e., volume) and its scientism, and their likely implications on HT. More specifically, the "suspicion" in the understanding, managing, and researching HT, which is growing under the full shadow of CT, was brought to the readers' attention. From where I stand, the search for and the launch of new tourism forms, whether humanistic, exotic, erotic, etc., appears like another neoliberal game for *touris(m)tification*, which has been in place for a long time. Seemingly, researchers, marketers, and tourist authorities are keen to take part in playing this game for the sake of enhancing profits and prestige. But as the Covid-19 pandemic shows, not all big-scale games are harmless! The CT is exploitative and as Butler notes, no forms of tourism is mutually exclusive with CT, which has turned into an open museum, in which all the feminine subjects and objects are thought, designed, transformed and exhibited to serve men, in response to the quest for devouring desire of masculinity (Franklin & Crang, 2001; Tribe, 2006; Tribe 2010; Tribe & Airey, 2007). That is, any new forms of tourism which are pursuing beauty and pleasure, and which are trying to flourish under the ideological shadow of CT cannot be pronounced entirely humane.

Producing another mainstream tourism product in the name of humanism in order to sustain the profits under the shadow of CT will not be humanistic unless ideological problems surrounding the CT and its research paradigm are recognized and addressed. To start with, taking a narrow perspective and relying particularly on centralized theories and research paradigms of the CT for understanding, managing, and researching HT is very disputable. The current research paradigm in CT values explanation and prediction more than it values understanding. We need a humanistic research paradigm that explains how one connects with something other than himself and how one can translate tourism experiences into values for a just world. Humanistic frameworks are needed to understand tourists who are temporarily living in a social experiment laboratory, testing their new skills to see the results of new relationships, the results of their many new identities, the results of their unlimited travels for personal growth through delayed gratification. Unless we shift focus from prediction for *hedonism* to understanding of *eudemonia*, the HT research is likely to produce the often misleading and frustrating results of the CT research (see Ioannidis, 2005; Kollat, Blackwell, Engel, 1972; Rossiter, 2011).

The HT, though claimed to be different, is being conceptualized similar to the CT. It is defined as a reaction to external stimuli and for consuming humanistic social, cultural, environmental tourism resources (Yüksel, 2017). In contrast to this business-driven view, HT should be conceptualized as an intrinsic, transformational experience for personal development. Therefore, the fetishist belief that exists in CT literature stating that the tourism movement is an economic behavior should be revisited. Taking HT from an economic point of view would be a very incomplete and also misleading picture (Franklin & Crang, 2001; Tribe, 2006; Tribe 2010; Tribe & Airey, 2007) since tourism is an experiential investment for personal development. As long as tourism activity involves any form of monetary exchange, it may not be reasonable to talk about pure humanity at all. To make economic profits without exploitation seems to be a fantasy. In addition, the ideological profit – influence or congruence by one culture over the people of the underdeveloped world – constitutes a much tougher obstacle for HT to flourish. In short, if tourists continue to be oblivious and bargain for lower prices, ask for free items despite the poverty surrounding them, how can we expect that HT will wipe out problems associated with CT? (Errington & Gewertz, 1989; O'Rouke, 1987).

Tourism has long changed from being a mortal and a short ritual to a modality through which ever-increasing modern life is being organized (Franklin & Crang, 2001; Tribe, 2006; Tribe 2010; Tribe & Airey, 2007). Hence, the HT should not be seen as a specialized consumer product or a consumption type. It is an experience with the potential to lead a more meaningful life in a just world, "for self-development, personal change and growth" (Botterill, 1989, p. 282). It enables "... *the individual tourist to capitalize on the experience of being a tourist, or any other leisure persona, as a part of becoming a person*" (Botterill, 1989, ibid.) As stated earlier, the HT can be better understood from an intrinsic, transformational perspective since humanism means transformation from seeking pleasure to seeking meaning. And in this transformational journey, whether abstract or concrete, facts and phenomena are transformed into an individualized reality. Several means that are open to the individual's creation, alteration, and construction are incorporated in this transformation. Hence, the HT can resemble a compass that is required to set off on the right direction for achieving "self-actualization in a socially responsible way".

In my opinion, humanistic tourists are distinctly different. This is because the basic need of humanistic tourists is not just the biological rooted deficiency needs but the "being needs" including, character growth, maturation, development, and helping others (Maslow, 1967). The humanistic tourist transforms not only herself but also her immediate surroundings by converting the time into valuable, dignified experiences and then translating these experiences into values (Yüksel & Yanık, 2014). In short, the HT should be in pursuit of a return to the essence, to the self and *eudemonia*. In this sense, humanistic tourists are expected to look at things with a second eye, immerse in the details of the moral life rather than the details of the artificial structure erected for the tourist business (Franklin & Crang, 2001; Tribe, 2006; Tribe 2010; Tribe & Airey, 2007).

As aforementioned, the HT appears to be flourishing under the full shadow of CT and its scientism. However, taking a narrow and biased perspective and relying particularly on centralized theories and research paradigms of CT for managing and researching HT would be disputable. That is, assuming to reach an understanding of HT with conceptual and methodological tools designed merely for "prediction" of tourist behavior in the CT is a fantasy. CT promotes large-scale business but the HT should thrive on by remaining small-scale. As the Covid-19 pandemic shows, the future is not in big-scale business, say hotels with thousands of rooms, but in small, responsible, boutique style of scale. In sum, in order to pave the way for a big change in consciousness of tourism, we need to understand (1) Why the tourism industry seems to have started to be more concerned with humanistic issues (moral and spiritual world)? (2) How so-called "humanism" will bind up the wounds stemmed from the destructive structure of CT, which continues to enlarge despite the hosts' voiced sufferings from pollutions, crowds, environmental problems, segregations, etc.? (3) What are the most important features that separate humanistic tourists and HT from other types, say CT? (4) Should humanism be conceptualized as a psychological *pull* factor traveling for certain parts of destinations or a psychological *push* factor leading to a meaningful tourist experience? (5) To what extent and how current scientism will respect and recognize moral obligations and effects of research(er) actions, as well as, humanistic and persona-centered understanding of tourist-host experiences? (6) How will the tourism industry realize the benefits of HT without focusing on "profits?" and (7) how can academia prevent HT from becoming another mainstream market product, serving only to business purposes?

Note

1 Before delving into the chapter, I would like to stress that the intention behind this critique is not renouncing past research on tourism but to provide an alternative perspective in a socially changing world. The readers are advised to bear in mind that the results of the studies referred to in explaining whether the vacation is a trauma may be dependent on several other variables including the travel type, duration, condition of health, travel propensity, etc.

References

Benavides, D. D. (2001). *Tourism in the Least Developed Countries*. Madrid: World Tourism Organization.

Bigsten, A., & Shimeles, A. (2007). Can Africa reduce poverty by half by 2015? The case for a pro-poor growth strategy. *Development Policy Review*, *25*, 147–166.

Bizzari, C. (2016). Opportunities and costs of tourism for new humanism. *Miscelenea Geographica – Regional Studies on Development*, *20*(2), 13–18.

Botterill, T. D. (1989). Humanistic tourism? Personal constructions of a tourist: Sam visits Japan. *Leisure Studies*, *8*(3), 281–293. doi: 10.1080/02614368900390281.

Buchanan, R. D., (1998). Beyond positivism: Humanistic perspectives on theory and research in health education. *Health and Education Research: Theory and Practice*, *13*(3), 439–450.

Buhalis, D., (1999). Limits of tourism development in peripheral destinations: Problems and challenges. *Tourism Management*, *20*(2), 183–185.

Burns, Peter & Novelli, Marina, (2008). *Tourism development: Growth, myths, and inequalities*. Wallingford, Oxfordshire, UK: CABI.

Butler, W. R., (1999). Sustainable tourism: A state-of-the-art review. *Tourism Geographies*, *1*(1), 7–25. doi: 10.1080/14616689908721291.
Cater, E., (1995). Environmental contradictions in sustainable tourism. *The Geographical Journal*, *161*(1), 21–28.
Caton, K., (2012). Taking the moral turn in tourism studies. *Annals of Tourism Research*, *39*(4), 1906–1928.
De Bloom, J., Geurts, S., &ad Kompier, M., (2010). Vacation from work as prototypical recovery opportunity. *Gedrag & Organisatie*, *23*(4), 333–349.
De Bloom, J., Geurts, S. A. E., Taris, T. W., Sonnentag, S., De Weerth, C. & Kompier, M. A. J., (2010). Effects of vacation from work on health and well-being: Lots of fun, quickly gone. *Work & Stress: An International Journal of Work, Health & Organisations*, *24*(2), 196–216.
De Bloom, J., Kompier, M. A. J., Geurts, S. A. E., De Weerth, C., Taris, T. & Sonnentag, S., (2009). Do we recover from vacation? Meta-analysis of vacation effects on health and well-being. *Journal of Occupational Health*, *51*(1), 13–25.
Eden, D., (1990). Acute and chronic job stress, strain, and vacation relief. *Organizational Behavior and Human Decision Processes*, *45*, 175–193.
Eden, D., (2001). Vacations and other respites: Studying stress on and off the job. In C. L. Cooper & I. T. Robertson (Eds.), *International review of industrial and organizational psychology* (Vol. 16, pp. 121–146). New York: Wiley.
Egresi, I., (2018). Tourists go home! – Tourism overcrowding and "tourismophobia" In European cities
(can tourists and residents still co-habitate in the city?) Paper presented at [Co] habitation Tactics. Imagining future spaces in architecture, city, and landscape. *TAW2018 International Scientific Conference*, 20–23 September 2018, POLIS University.
ENP. (2014). *Post-holidays and the broken promise effect – An EAP's perspective page* 2. Insights from the Shepell·fgi Research Group (2007). Post-holidays and the broken promise effect – an EAP's perspective. Retrieved from http://www.shepellfgi.com/EN-CA/AboutUs/News/Research%20Report/pdf/PostHolidayResearchReportENGFinal.pdf.
Errington, f. & Gewertz, D., (1989). Cannibal tours. 1987. A film by Dennis O'Rourke. *American Antropologist*, *9*, 274–275.
Franklin, A. & Crang, M., (2001). The trouble with tourism and travel theory. *Tourist Studies*, *1*(1), 5–22.
Fritz, C. & Sonnentag, S., (2005). Recovery, health, and job performance: Effects of weekend experiences. *Journal of Occupational Health Psychology*, *10*(3), 187–199.
Fritz, C. & Sonnentag, S., (2006). Recovery, well-being and performance related outcomes: The role of workload and vacation experiences. *Journal of Applied Psychology*, *91*(4), 936–945.
Fouberg, E. H., Murphy, A. B. & deBlij, H. J., (2015). *Human geography: People, place and culture* (11th ed.). Wiley.
Ghatak, S., (1995). *Introduction to development economics* (3rd ed.). London: Routledge.

Goytia, A. P. & de la Rica, A., (2014). Personal experience tourism: A post modern understanding. In R. Sharpley & P. R. Stone (Eds.), *Contemporary tourist experience: Concepts and consequences.* London: Routledge.

Gump, B. B. & Matthews, K. A. (2000). Are vacations good for your health? The 9-year mortality experience after the multiple risk factor intervention trial. *Psychosomatic Medicine*, *62*(5), 608–612.

Halevi, G. & Bar-Ilan, J., (2016). Post retraction citations in context. In *BIRNDL 2016 bibliometric-enhanced information retrieval and natural language processing for digital libraries* (pp. 23–29). Newark: CEUR.

Hartig, T., Catalano, R., Ong, M. & Syme, S. L., (2013). Vacation, collective restoration, and mental health in a population. *Society and Mental Health*, *3*(3), 221–226.

Hirschman, E. C., (1986). Humanistic inquiry in marketing research: Philosophy, method, and criteria. *Journal of Marketing Research*, *23*(3), 237–249.

Hoopes, L. L. & Lounsbury, J. W., (1989). An investigation of life satisfaction following a vacation: A domain-specific approach. *Journal of Community Psychology*, *17*(3), 129–140.

Hundt, A., (1996). Impact of tourism development on the economy and health of Third World Nations. *Journal of Travel Med*, *3*(2), 107–112.

Ioannidis, JPA (2005). Why most published research findings are false. *PLoS Medicine*, 2(8), e124.

Jover, J. & Díaz-Parra, I., (2019). Gentrification, transnational gentrification and touristification in Seville, Spain. *Urban Studies.* August, 1–16. doi: 10.1177/0042098019857585.

Koç, E. & Boz, H., (2014). Psychoneurobiochemistry of tourism marketing. *Tourism Management*, *44*(October), 140–148.

Kollat, D. T., Blackwell,, R. D. & Engel,, J. E. (1972). The current status of consumer behaviour research: Development during the 1968-1972 period. In Proceedings of the *3rd annual conference of the association of consumer research*, edited by M. Venkatesan (pp. 576–584).

Kontogeorgopoulos, N., (2004). Conventional tourism and ecotourism in Phuket, Thailand: Conflicting paradigms or symbiotic partners. *Journal of Ecotourism* 3(2), 87–108.

Kop, W. J., Vingerhoets, A., Kruithof, G.-J. & Gottdiener, J. S., (2003). Risk factors for myocardial infarction during vacation travel. *Psychosomatic Medicine*, *65*(3), 396–401.

Krippendorf, Jost, (1987). *The holiday makers: Understanding the impact of leisure and travel.* London: Heinemann.

Lewis-Beck, M. S., Bryman, A. & Liao, T. F., (2004). *The SAGE encyclopedia of social science research methods.* London, Thousand Oaks: Sage Publications, Incorporated.

Maslow, A.H., (1967). A theory of metamotivation. *Journal of Humanistic Psychology*, 7(2), 93–127. Reprinted as chapter 23 in The farther reaches of human nature, 1971.

Minquez, C., Pineria, J. M. & Tabales, A., (2019). Social vulnerability and touristification of historic center. *Sustainability*, *11*(16), 4478. doi:10.3390/su11164478.

Nawijn, J., (2011a). Determinants of daily happiness on vacation. *Journal of Travel Research*, *50*(5), 559–566.

Nawijn, J., (2011b). Happiness through vacationing: Just a temporary boost or long- term benefits? *Journal of happiness studies*, *12*(4), 651–665.

Nawijn, J., De Bloom, J. & ve Geurts, S., (2013a). Pre-vacation time: Blessing or burden? *Leisure Sciences*, *35*(1), 33–44.

Nawijn, J., Marchand, M., Veenhoven, R. & ve Vingerhoets, A., (2010). Vacationers happier, but most not happier after a holiday. *Applied Research in Quality of Life*, *5*(1), 35–47.

Nawijn, J., Mitas, O., Lin, Y. & ve Kerstetter, D., (2013b). How do we feel on vacation? A closer look at how emotions change over the course of a trip. *Journal of Travel Research*, *52*(2), 265–274.

O'Rouke, D., (1987). *Cannibal tours: Produced by Dennis O'Rourke*. Direct Cinema Limited (PO Box 69799, Los Angeles CA 90069, USA) 1987, 77 minutes, 16 mm film and 1/2" video format, $995 (16 mm), $350 (video).

Palmer, C., (1994). Tourism and colonialism: The experience of the Bahamas. *Annals of Tourism Research*, *21*(4), 792–811.

Pearce, P. L., (1981). Environment shock: A study of tourists' reactions to two tropical islands. *Journal of Applied Social Psychology*, *11*(3), 268–280.

Pritchard, A., (2014). Gender and feminist perspectives in tourism research. *A Tourism Companion*, 314–324.

Restak, R., (2001). *The secret life of the brain*. Joseph Henry Press.

Robinson, J., (2004). Squaring the circle? Some thoughts on the idea of sustainable development. *Ecological Economics*, *48*, 369–384.

Rossiter, J. R., (2011). *Measurement for the social sciences: The C-OAR-SE method and why it must replace psychometrics*. New York: Springer.

Ryan, C., (1991). *Recreational tourism: A social science approach*. London: Routledge.

Sequera, J. & Nofre, J., (2018). Debates shaken not stirred: New debates on touristification and the limits of gentrification. *City*, 1–13, https://doi.org/10.1080/13604813.2018.1548819.

Sharpley, R., (2009). *Tourism development and the environment: Beyond sustainability?* UK: Eartscan.

Simkin, M & Roychowdhury, V., (2002). Read before you cite!. *Complex Systems*, *14*, 269.

Sonnentag, S., (2003). Recovery, work engagement, and proactive behavior. A new look at the interface between nonwork and work. *Journal of Applied Psychology*, *88*(3), 518–528.

Sorell, T., (1991). *Scientism: Philosophy and the infatuation with science*. New York: Routledge.

Strauss-Blasche, G., Ekmekcioglu, C. & Marktl, W., (2000). Does vacation enable recuperation? Changes in well-being associated with time away from work. *Occupational Medicine*, *50*(3), 167–172.

Strauss-Blasche, G., Ekmekcioglu, C. & Marktl, W., (2002). Moderating effects of vacation on reactions to work and domestic stress. *Leisure Sciences*, *24*(2), 237–249.

Tarumi, K. & Hagihara,, A. (1999). An inquiry into the causal relationship among leisure vacation, depression, and absence from work. *Journal of UOEH, 21*(4), 289–307.

Tonelo, D., Providência, R. & Gonçalves, L., (2013). Holiday heart syndrome revisited after 34 years. *Arg Bras Cardiol, 10*(12), 183–189. Retrieved from https://www.ncbi.nlm.nih.gov/pmc/articles/PMC3998158/.

Tosun, C., (2000). Limits to community participation in the tourism development process in developing countries. *Tourism Management, 21*(6), 613–633.

Tribe, J., (2006). The truth about tourism. *Annals of Tourism Research, 33*(2), 1–20.

Tribe, J., (2010), Tribes, territories and networks in the tourism academy. *Annals of Tourism Research, 37*(1), 7–33.

Tribe, J. & Airey, D., (2007). *Developments in tourism research. New directions, challenges and applications*. New York: Elsevier Science.

Van Heck, G. L. & Vingerhoets, J. J. M., (2007). Leisure sickness: A biopsychosocial perspective. *Psychological Topics, 16*(2), 87–200.

Vingerhoets, A. J. J. M., van Huijgevoort, M. & van Heck, G. L., (2002). Leisure sickness: A pilot study on its prevalence, phenomenology, and background. *Psychotherapy and Psychosomatics, 71*(6), 311–317.

Wamoth, A., (1998). Humanistic psychology and humanistic social science. *Humanity & Society, 22*(3),313–319.

Westman, M. & Eden, D., (1997). Effects of a respite from work on burnout: Vacation relief and fade out. *Journal of Applied Psychology, 82*(4), 516–527.

Westman, M. & Etzion, D., (2001). The impact of vacation and job stress on burnout and absenteeism. *Psychology and Health, 16(5), 595–606.*

Yüksel, A., (2016). Tatil, Sağlık ve iş performansı: Tarvma ya da tedavi? *Soid, 13*(2), 123–120.

Yüksel, A., (2017). A critique of "Response Bias" in the tourism, travel and hospitality research. *Tourism Management, 59*(April), 376–384.

Yüksel, A., (2019a). *Aykırı. Bilimsel Araştırmada yanılsamalar (Divergent: Delussions in scientific research).* Ankara: Detay Yayıncılık.

Yüksel, A., (2019b). Mystery of spiritual intelligence: Predictions, prophecies and possibilities. In Koç, E., *Emotional Intelligence in tourism and hospitality* (pp. 134–147). Ankara: Cabi Publishing.

Yüksel, A. & Yanık, A., (2014). Co-creation of value on social media: How?. In Prebensesn, N. Chen, J. & Uysal, M. (Eds.), *Creating experience, value and tourism*. Oxfordshire: Cabi Publishing.

Zanardi, C., (2019). Complexity-cage: Toward a critical deconstruction of "tourism-fobia". *Urban Transcripts Journal*, 2, 1. http://journal.urbantranscripts.org/article/complexity-cage-toward-critical-deconstruction-tourism-phobia-clara-zanardi/.

Zhou, J., (2017). Study on the development of humanistic tourism resources in the development of tourism industry. Advances in social science, education and humanities research, Vol. *185. 6th international conference on social science, education and humanities research* (SSEHR 2017).

3 Voluntourism, Humanistic Management, and Relational Goods

Plinio Limata and Giorgia Nigri

1 Introduction

The non-instrumental relational positives defined by Smith (1894) and the human relational nature outlined by the civil economy tradition (Bruni, 2014) have been central pillars in economic thought. Scholars have involved relational analysis in studying what creates happiness for individuals and the utility deriving from encounters while consuming a good, rather than the utility of the good itself. Defined as non-material goods, which are intrinsically linked to interaction and reciprocity, relational goods are now regarded as a multidisciplinary concept.

This notion can be adopted while analyzing the impact deriving from voluntourism because there is still a significant challenge in assessing the qualitative impact (especially on the host community) of this activity, given the multi-perspective character, the variety of stakeholders involved, and the different scenarios in which it takes place (Lupoli & Morse, 2014). Although the debate is still ongoing on what should be the right and comprehensive definition (Limata, 2017), the concept of voluntourism comprises both volunteerism and tourism (Stainton, 2016), being commonly known as a type of holiday in which the vacationer works as a volunteer to help people in the visited host country (e.g., Brown, 2005; McGehee & Santos, 2005; Wearing, 2011).

As it will be shown, the sector has grown enormously in recent years, and this has led to a possible change in its nature and to the development of numerous critical analyzes which have weakened the positive and uncritical studies conducted in the early days of this particular kind of tourism. However, the relational nature of this activity is what drives the application of the theory of relational goods to this niche tourism market, in order to foster a better comprehension of it and push for a new conceptual framework to analyze it.

To follow this path, we decided to structure the article as follows. First, we briefly explain the methodology we used to carry out the literature review. Second, we present voluntourism in light of the relational goods theory approach. We then analyze results under the humanistic management umbrella and finally introduce some practical implications and our conclusions.

2 Literature Review: Methodology

Research on the literature of both topics has been carried out in order to investigate the possibility of applying the concept of relational goods and humanistic management to voluntourism. A literature review was developed following Vom Brocke et al. (2009) and Webster and Watson (2002), as shown in Figure 3.1.

We followed the three-stage procedure: planning, execution, and reporting (Fink, 2005; Tranfield, Denyer, & Smart, 2003). In the planning stage, we defined the specific aims of the research, and the key data sources. We focused on scholarly literature featuring the lens of civil economy. We queried scientific databases, such as JSTOR, Scopus, SSRN, Science Direct, Ideas, and G-Scholar, as they are the most comprehensive of peer-reviewed journals in the management, organization and social science fields (Nigri & Michelini, 2019), using the following keywords: volunteer tourism, voluntourism, relational goods, humanistic management, identifying 45 articles.

The execution phase was then carried out in four steps. (1) The first layer of analysis referred to titles, abstracts, and keywords, to proceed to the exclusion of those articles not directly focused on the issues investigated or mentioning them in a superficial context; (2) the second step was a snowball analysis, (3) which was performed to extend the coverage of the research. This additional step was conducted in order to process 21 additional articles. (4) Lastly, we selected 66 articles, and a classification was performed that divided the materials into themes regarding the specific topic concerned.

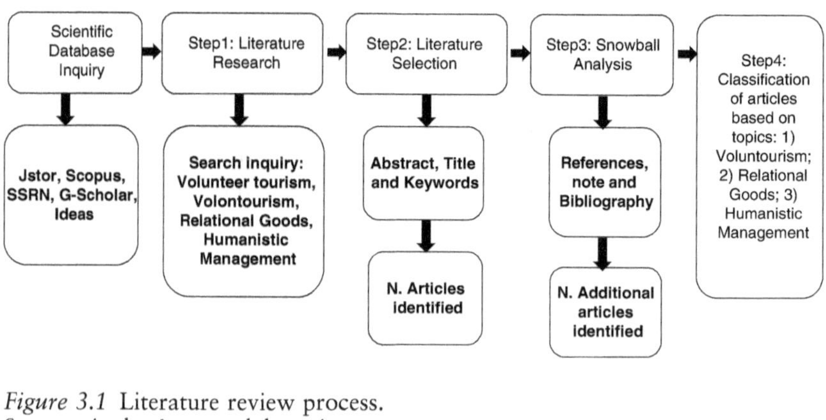

Figure 3.1 Literature review process.
Source: Author's own elaboration.

2.1 Relational Goods

As mentioned by Bruni (2013), relational goods as a concept were introduced simultaneously by four authors: the philosopher Martha Nussbaum (1986), the sociologist Pierpaolo Donati (1986), and the economists Benedetto Gui (1987) and Carole Uhlaner (1989). What is common to the several definitions is the interpretation of the relationship as a good: the relationship between subjects is not seen as a means, but the end itself (Bruni, 2011). Relational goods, rooted in the Aristotelian tradition (Bruni, 2010), indicate the affective/expressive, non-instrumental side of interpersonal relationships (Becchetti, Pelloni, & Rossetti, 2008). They cannot be produced, consumed, or acquired by a single individual because they depend on the interaction with others and are exclusively enjoyed if shared (Bruni, 2008). In literature, they are described as "emergent effects of an action, not the effect of the choices of the actor, nor of the environment but the product of concrete relationships, which can change the will of the actors themselves" (Donati, 2005, cited in Bruni, 2011).

Relational goods are thus peculiar intangible outputs of an effective and communicative nature (Gui, 2000) that are produced through social interactions (Becchetti, Degli Antoni, & Faillo, 2010a; Uhlaner, 1989). Social interactions differ from interpersonal relationships because, in the latter, the identity of the subjects involved is constitutive of the relationship itself, whereas social interactions can be anonymous and impersonal (Zamagni, 2006). Mentioning Becchetti, Degli Antoni, and Faillo (2010b), it is possible to affirm that relational goods have three main characteristics: (1) they are a subset of local public goods because they are non-rival and non-exclusive but only about the individuals who participate in their production; (2) contributions to their production depend on a mutual agreement – goodwill, in this case, is essential for their production as they cannot be imposed; (3) their value depends on the characteristics of people sharing the goods and is increased by reciprocal fellow feeling (Becchetti et al., 2010b).

According to Gui (2000), much economic activity takes place not through market exchanges but through market *encounters*, where relatedness plays a crucial role, having the potential of bringing to different outcomes. Therefore, it is the relationship itself that constitutes the "good" in relational goods – the dimension of reciprocity is foundational (Bruni, 2010). Identity, reciprocity, simultaneity, gratuitousness, emerging facts, and goods are all key characteristics of interpersonal relationships (Bruni, 2011). Furthermore, the identity of the other person is essential for the value and sometimes the existence itself of the relational good (Bruni, 2010).

Finally, relational goods can be divided (Donati, 2003, cited in Bruni & Zarri, 2007) into two large families: primary and secondary goods.

Although non-primary relational goods can be considered as an additive outcome to the other components produced by the interaction-encounter relationship, in primary relational goods, the relational component of the encounter cannot be eliminated without destroying the value of the good itself. It must be noted that primary relational goods cannot be transformed into money because assigning them a monetary value would distort their worth (Bruni & Zarri, 2007).

Relational goods are abundant in social life as they are produced and consumed simultaneously by those interacting in the relationship, producing social capital. Social factors, social environment, social status, and the ability to construct and enjoy meaningful and satisfactory relations with other people are at the basis of wellbeing (Antoci, Sacco, & Vanin, 2007). Time spent with the family – parents and other relatives – has the largest effect on life satisfaction and individual happiness, followed by activities where intrinsic motivation plays a prominent role – such as charity, church, and arts – and activities that imply a joint effort for a common cause – such as unions, parties, environmental organizations (Becchetti et al., 2008; Bruni & Stanca, 2008). The relational component of participation in voluntary organizations is so important that its effect on life satisfaction is quantitatively similar to that of moving up by one decile in the income scale (Bruni & Sugden, 2006).

The time and effort devoted to the production and consumption of relational goods are indeed much lower today than in the past (Bruni & Sugden, 2006). However, voluntourism poses great attention to relational and social factors.

2.2 *Voluntourism*

In the literature, there is a growing interest in voluntourism, volunteerism, and volunteer tourism (Limata, 2017). “Volunteer tourism is a term that has been used to describe a wide range of tourist behaviors and tourism products and services and is now one of the fastest-growing forms of alternative tourism” (Lyons & Wearing, 2008, p. 6). As Stainton (2016) synthesizes, the concept comprises both volunteerism and tourism, which, despite beginning as a micro-niche form of tourism, it can now be found in virtually every industry (Wearing & McGehee, 2013), even in mass tourism (Butcher & Smith, 2010). It has come to be considered as an extension of ecotourism (Gray & Campbell, 2007; Wearing, 2011; Wearing & McGehee, 2013), as new tourism (Poon, 1993), as niche tourism (Novelli, 2005), or as new moral tourism (Butcher, 2003, 2005). Today, volunteer tourism is associated with social tourism, charity tourism, moral tourism, and serious tourism – all experiences which foresee voluntourists being engaged in working on projects within local communities (Suanpang, Srisuksai, & Tansutichon, 2018). Moreover, as highlighted by Tomazos and Cooper (2011), it can be defined as a "volunteer vacation" (McMillon, Cutchins, & Geissinger, 2006), "mini-mission" (Brown, 2005), "mission-lite", "pro-poor tourism" (Ashley, & Roe, 2002; Hall, 2007), "altruistic tourism" (Singh, 2014), and "service-based vacation" (Ellis, 2003). The term "voluntourism" (The Guardian, 2007), for the sake of simplification, will be used in this chapter to encompass all the readings cited earlier.

In his book Volunteer Tourism – Experiences That Make a Difference (2001), which according to Benson (2011) acted as the catalyst of the literature that followed, Wearing (2011, p. 1) defines voluntourists as those who "volunteer in an organized way to undertake holidays that may involve the aiding or alleviating of the material poverty of some groups of society, the restoration of certain environments, or research into aspects of society and environment". Voluntourism could, therefore, be seen as a form of alternative tourism, capable of benefitting the volunteer and the host community, having the potential to elevate both the giver and the receiver (Brown, 2005; Wearing, 2011).

Wearing's attempt to define the topic brought to several other definitions, first and foremost the one provided by practitioners: "a seamlessly integrated combination of voluntary service to a destination along with the best, traditional elements of travel- arts, culture, geography, and history- in that destination" (VolunTourism International, 1999). Successively, McGehee and Santos (2005, p. 760) define volunteer tourism as "utilizing discretionary time and income to travel out of the sphere of regular activity to assist others in need". Basically, voluntourism definitions consider two core elements: leisure time and work (volunteering) for the benefit of others – two aspects which could be easily viewed as contradictory but which, on the contrary, eliminate boundaries. A characteristic of postmodernism is the blurring and disappearance of boundaries, "in volunteering and tourism, the combination of work and leisure is not a contradiction, but a rational match" (Benson, 2011, p. 215).

Alternatively, it can be said that we are experiencing a reconfiguration of the relationship of the individual to society, where individual identity becomes a key factor of political change because contemporary society is no longer bound to customs and traditions (Giddens, 1994). When people are no longer guided by traditions such as class, religion, or family, life politics represents the attempt to create morally justifiable lifestyles that bridge the gap between individuals and social change. This aspect is evident when analyzing the origins of voluntourism. Butcher and Smith (2010) speak of modern development volunteering emerging after the Second World War, whereas Benson (2011) pinpoints its roots in the 19th century. Examples include military volunteers in Latin America in the 1980s and Spain in the 1930s signing up to fight for independence (Simpsons, 2004), or civil service projects in Europe in the 1920s set up by Service Civil International (Tomazos & Butler, 2012).

Indeed, it must be noted that the experiences associated with voluntourism seem to confirm the importance of the engagement of individuals based on an intrinsic motivational drive. Moreover, what seems to be foundational of this activity is that the action seems always orientated toward the others and their needs, therefore being deeply relational. This particular aspect will be addressed later.

3 Voluntourism in Light of the Humanistic Management Notion

Humanistic management serves the societies in which businesses operate, increasing the quality of life (Melé, 2003; Spitzeck, Pirson, Amann, Khan, & von Kimakowitz, 2009). It, therefore, calls for a transition from a purely economic orientation toward a more human-centric approach to alleviate human suffering and to contribute to life conduciveness (Ulrich, 2009). Effective mechanisms should be put in place to foster human dignity, as human dignity has a value (Kant, 1997).

Scholars and practitioners frequently refer to voluntourism as "making a difference" (Wearing, 2011), referring not only to the difference volunteers could/would make in host communities, but "to the deep, personal transformation voluntourism can bring about in the volunteer" (Wearing, 2011, p. 4). Voluntourism can support the development of solidarity and tolerance among volunteers and the community, enabling the first to gain experiences and skills to ensure their personal as well as professional growth (Pompurová, Marčeková, Šebová, Sokolová, & Žofaj, 2018). Meadows mentions that the highest leverage point to intervene in a system is the mindset or paradigm out of which the system arises (Meadows, 1997). Barnard (1938) argued that inquiry into human nature was an indispensable prerequisite of social systems, which required human cooperation in order to achieve success. Another approach of humanistic management stresses the importance of social systems for realizing personal values (Bruni, 2009Mion 2011; Melé, 2011; Mion & Adaui, 2011) and considers the relation of humanism and business and economics in different religious and cultural traditions such as the Christian social tradition. Thus, transforming consciousness is the most effective tool for unlocking local and global change (Tsao & Laszlo, 2019), and voluntourism experiences give this opportunity in host communities. The benefits reported include raised consciousness, increased interest in activism and increased prosocial values, compassion, and perspective-taking (Bailey & Russell, 2012; Wearing, 2011). The concern for persons and human aspects is thus at the heart of voluntourism and is a construct of humanistic management. Humanism stresses the dignity and worth of human beings and the search of good for them (Loza, Cristian, & Habisch, 2013). It is oriented not only to obtain results through people but also, and above all, toward people themselves, showing care for their flourishing and wellbeing – voluntourism can "facilitate positive change" and help voluntourists discover their "role in the universe and society" (Germann Molz, 2015).

As Bruni and Sugden (2006) asked themselves, "what is the nature of the relationship between trading partners in a market? Does this relationship, as perceived by the partners themselves, have any social or moral content?" These questions become salient whenever there is a perception that the domain of the market is expanding, as in the voluntourism sector. Such a perception prompts people to ask whether something valuable is lost when non-market modes of interaction are replaced by market ones (Bruni & Sugden, 2006). For example, as mentioned earlier, "membership in a voluntary organization is associated with a statistically significant increase in life satisfaction, ceteris paribus" (Bruni & Stanca, 2008, p. 517).

As highlighted by critics, significant increases in the number of commercial operators have changed the face of the industry (Wearing & McGehee, 2013). Today it could be said that volunteer tourism is subject to the same market rules as conventional tourism (Skoczylas, 2012). The handling agencies can range from a locally based non-profit to a multinational commercially-run organization (Raymond, 2008, p. 49). The marketplace is ever-evolving, with new businesses entering the market, ranging from charitable or non-profit (Brown, 2005) to projects funded by large institutions such as the World Bank (Wearing & McGehee, 2013) and traditional tour operators (Benson & Henderson, 2011; Stainton, 2016Benson 2011).

However, as Bailey and Russell (2012) mention, many researchers confirm that volunteer tourists are a unique clientele who are actively seeking interaction and a deeper understanding of diverse cultures. In their analysis, Vodopivec and Jaffe (2011) recall the concept of the volunteer encounter built on Cohen's (1984) "tourist encounter", which can be described as a transitory encounter in which actors are not oriented toward the maintenance of a continued relationship.

However, as Crouch argues, "tourism becomes validated in human practice in relation to knowledge. Knowledge is constructed through encounters, and space is important in informing this knowledge" (Crouch, 2002, p. 205 cited in Sin, 2009). Thus, the duration of the experience is not relevant; what matters is the intensity of it. Bearing this in mind, there is no need to understand the motivations of the volunteer tourist because a distinctive feature of relational goods is that the consumption of a subject increases with the increase of the time that the subject dedicates to socialization and with the socialization effort expressed by other subjects (Bruni & Zamagni, 2004; Wearing & McGehee, 2013).

Whatever the motivation, the encounter among the two parties will take place, and this could lead to the development of sincere fellow-feelings among voluntourists and host community individuals. Therefore, it seems possible to assess that one of the essential aspects of voluntourism is the ability to construct and enhance people's networks and shared knowledge, together with the ability to amplify relational activities that lie at the basis of wellbeing and thus utilize a humanistic management approach.

4 Critical Aspects of Voluntourism

It is not until the last 30 years that voluntourism has become popular, reaching the pattern it has today, in opposition to mass travel (Skoczylas, 2012). Young (2008) considers volunteer tourism as an expanding sector of the tourism industry in both developed and developing countries. It is becoming an increasingly popular form of leisure travel for people of all ages (Bailey & Russell, 2012) typically involving volunteers from high-income countries traveling to low- and middle-income countries to assist in service-related developmental activities. These experiences may vary in form, duration, and objectives; a common and potential goal of the short-term volunteerism is often seen as filling gaps in local services in the short-term while building sustainable systems to promote self-sufficient communities in the long-term (Loiseau, Sibbad, Raman, Darren, Loh, & Dimaras, 2016). Although it has mainly a positive connotation, especially under the humanistic management perspective, it also has some severe limitations that we will discuss later.

4.1 Commodification and Concerns

Voluntourism is generally considered reciprocally beneficial for both the tourists and the host community residents (Woosnam & Lee, 2011), generating positive impacts and fostering a mutually beneficial relationship between host and guest (Lupoli & Morse, 2014; Sin, 2009). The first academic research in this field was based on these assumptions. Negative impacts went first unquestioned (Limata, 2017). This new form of niche tourism combining consumption and participation was considered by the World Tourism Organization as less intrusive and more beneficial to host communities because it was meant to guarantee a just positive impact. The growth of the market sector though attracted several commercial organizations which led to a shift in the opposite direction. Voluntourism went from being a de-commodified market to a commodified one, with the risk of converting cultural value to commercial value (Shepherd, 2002; Wearing & McGehee, 2013). If tourism in a free market economy represents the commercialization of the human need to travel, exploiting natural and cultural resources for profit accumulation (Wearing & McGehee, 2013), voluntourism then can be seen as "a particular, neoliberal form of development practice, in which development is not only privatized but can be packaged as a marketable commodity" (Vodopivec & Jaffe, 2011, p. 112). The services and experiences associated with volunteer tourism have thus moved from having "use-value" for the participant or the beneficiaries to having "exchange value" in the marketplace (Tomazos & Cooper, 2011).

This commodification process did not modify, at least apparently, the pillars of volunteer tourism – positive impacts to locals in host destinations and a reciprocal and mutually beneficial host-guest relationship in tourist destinations (Sin, 2010). However, a critical need for research to provide a firm foundation for a deeper understanding of volunteer tourism, in both its positive and negative aspects, emerged (Sin, 2009). As highlighted by Wearing and McGehee (2013), new critical studies emerged, and new kinds of potential threats were consequently analyzed. Volunteer tourism began to be seen as a new form of colonialism, reinforcing stereotypes and creating another layer of dependency between the developed and developing world, externalizing the development process and functioning more as a volunteer traveler's quest for career development than as a development tool (Fee & Mdee, 2011; Simpson 2004). Simpson (2004), in her research, focused on the negative aspects of these programs such as short-term vs. long-term care strategies, and the low level of skills required by the volunteers. Her claim was that voluntourism perpetuated a simplistic idea of development. It then was seen as a manifestation of ethnocentrism, a confirmation of stereotypes, and a source of conviction about the peculiarities of the inhabitants of the Third World (Skoczylas, 2012).

Nowadays, voluntourism is essentially a globally imbalanced phenomenon: 90% of aid-holidaying takes place in the "Global South" and a similar percentage of voluntourists are from Europe or North America (TRAM, 2008). Although voluntourism projects in developed countries are becoming more and more popular as volunteering in developing countries (Pompurová et al., 2018), still, voluntourism tends to reproduce the structural inequalities that the encounter is based on (Conran, 2011). Regarding this specific point, it is interesting to note that other researchers reached different conclusions, stating that voluntourism-generated direct contacts, between volunteers and inhabitants, disturb the stereotypes each have of one another (Raymond & Hall, 2008).

Other major concerns are related to the potential damages to the host community (Palacios, 2010), or the real possibility that volunteers could make a significant difference to the lives of the communities they claim to assist (Mustonen, 2005; Sin, 2010). Most of the literature and criticism of volunteer tourism comes from research on the experience of the volunteer rather than on the host community (Mdee & Emmott, 2008). Much of the research has been focused on volunteers and sending organizations, but more needs to be done when it comes to analyzing stakeholders and host organizations' impact. This issue was also highlighted by Sin (2010, p. 1), who affirms that "the lack of research and assessment of volunteer tourism in host-communities places much uncertainty on whether promised benefits are indeed realized". At the current moment, there is little consensus on the most appropriate methods for developing the appropriate indicators able to recognize the heterogeneous nature of stakeholders (Lupoli & Morse, 2014).

4.2 Motivations and Voluntourists

Although the impact on host communities needs to be better analyzed, when it comes to volunteers, much research has been done on investigating voluntourists' intrinsic motivations and personal impact. Research has been reaching several outcomes. Investigating on the voluntourism demand in Thai communities, Suanpang et al. (2018, p. 340) found that the most critical factors were "motivation" of tourists, their "attitude toward" volunteering, followed by their "expectations", and finally the "need" to help others.

Alternative studies (Lupoli & Morse, 2014) have identified other benefits that attract volunteer tourists, such as personal growth, first-hand work experience, cultural capital, language acquisition, career advancement, long-term friendships, and opportunities for adventure and self-reflection. Skoczylas (2012) and Tomazos and Butler (2012) identified three volunteer categories based on the relevant literature: 1) material/utilitarian; 2) solidary/affective/social; and 3) purposive/normative/altruistic.

Indeed, the desire to help others is a motivation but is just as important as the desire to gain hands-on experience and develop professional skills (Lyons, Hanley, Wearing, & Neil, 2012; Simpson, 2004). Through voluntourism, participants "create a 'personality package' which builds 'identity' and is instrumental to being competitive in the workplace" (Lyons et al., 2012, p. 370; Simpson, 2004). It can be considered as an effective way to signal on a curriculum vitae that the candidate has cultivated a range of soft skills (Germann Molz, 2015); in fact, voluntourists seem to be encouraged to focus on transforming themselves and pursuing "morally justifiable lifestyles" (Butcher & Smith, 2010, p. 30), entailing feelings, such as sadness, compassion, or intimacy (Germann Molz, 2015).

Synthesizing, although volunteer tourism is often conceptualized as prosocial behavior, the reasons for engaging in volunteer tourism are not always purely prosocial (Wong, Newton, & Newton, 2014). Volunteer tourists often have a multitude of motivations, and altruistic motivations are often not mutually exclusive with leisure seeking or self-development motivations (Sin, 2009). For young people, volunteering is becoming a popular way to travel, especially to developing countries (Vodopivec & Jaffe, 2011). Moreover, tourists are not only perceived as being ignorant of the problems faced by host communities but are also seen as contributing to them (Vodopivec & Jaffe, 2011). These contradictions emphasize a tension between the ascribed roles of "volunteer and tourist that is at the heart of voluntourism itself" (Lyons & Wearing, 2008, cited in Coghlan, 2015). This is made evident by the fact that voluntourists, combining volunteer work with tourism, distinguish themselves from tourists for two reasons: (1) authenticity of the experience and (2) work. Thus, what makes a true difference is the engagement level of travelers.

As remarked by Singh (2014), volunteer tourism develops feelings of global citizenship in the host as well as in the guest community (McGehee & Santos, 2005), and it improves cultural understanding in the world (Wearing, 2011). McGehee and Santos (2005) suggest that after cross-cultural experiences, volunteers feel a sense of global citizenship and become more involved in changing the world (Lee & Woosnam, 2010). Henderson and Presley (2003) contend that international volunteering may be a method of cultivating mutual understanding and building social networks across cultures (Bailey & Russell, 2012). McGehee and Santos (2005) recognized the ability of voluntourism to create connections (Coghlan, 2015). Concluding, as Conran (2011) remarks, although volunteer tourism has been identified as a neo-colonial development encounter, it has also been recognized as a platform for structural change. Volunteer tourist organizations bring together societal groups and people from different countries who work together to improve the lives of groups of people they would otherwise not have impacted (Tomazos & Cooper, 2011).

5 Discussion and Conclusions

Until nowadays, we can distinguish between two approaches that have been adopted to analyze the voluntourism sector: through the lens of Social Distance (Woosnam & Lee, 2011), or via the Social Exchange Theory (Paraskevaidis & Andriotis, 2017). Although the first refers to the grades and degrees of understanding and intimacy which characterize personal and social relations generally (Park, 1924, cited in Woosnam & Lee, 2011), Social Exchange theory analyzes the impact on host community and their interest in supporting tourism development when economic, environmental, and sociocultural impacts are perceived as positive (Paraskevaidis & Andriotis, 2017). In this theory, reciprocity is a concept that has been frequently taken into consideration in relation to the structure of social exchanges, which can be either direct or indirect. Direct forms of social exchanges are reciprocal exchanges, when "actor" contributions are separately performed, non-negotiated, and initiated by performing beneficial acts for another, whereas negotiated exchanges focus on the terms two actors negotiate to reach a mutually beneficial agreement (Paraskevaidis & Andriotis, 2017). The essential aspect of the relationship of reciprocity is that the transfers it generates are tied to human relationships: the objects of transactions are not separable from those who put them in place – in reciprocity the exchange ceases to be anonymous and impersonal as with the exchange of equivalents (Zamagni, 2006).

What emerges from the literature analyzed is that the voluntourism phenomenon could be better analyzed in light of the civil economy paradigm and the theory of relational goods. Regarding the first, as mentioned by Bruni and Sugden (2006), Antonio Genovesi treats reciprocal confidence as the fundamental "force of cohesion" in society; it is a precondition for friendship and humanity, as well as for commerce. The implication is that commercial relations are similar to interpersonal relations in other domains of civil life. Market relationships are understood by participants as relationships of mutual assistance, in which people make themselves useful to one another. Market relations are not different from other relations of civil society. The fundamental characteristic of market relations is reciprocity, and reciprocity is the governing principle of civil society.

In contrast, when it comes to the theory of relational goods, behavioral and experimental economics have offered important contributions indicating that interpersonal dimensions, such as "trust, fairness, and reciprocity, are relevant for economic decisions" (Bruni, 2008, p. 508). Consuming relational goods has been proved to have a direct impact also on variables such as trust and trustworthiness that are key elements for socio-economic development (Becchetti et al., 2010b). Time spent in producing and consuming relational goods is shown to generate significant and positive effects on life satisfaction (Becchetti et al., 2008), but as research has demonstrated, the time and the effort devoted to production and consumption of relational goods are decreasing in market societies (Bruni, 2010). Voluntourism presents the peculiarities of activity capable of reversing this tendency, or at least of building and enhancing people networks and sharing knowledge.

Although Bruni and Stanca (2008) argue that the absence of instrumental motivations is a key element affecting the quality of relational goods, reciprocity seems to be at the heart of the relations that are built among voluntourists and individuals in host communities. Recalling the distinction between social interactions and interpersonal ones offered by Zamagni (2006), we argue that, although the relations among voluntourists and service providers could be read as social interaction, relations among voluntourists and host community individuals should be read as interpersonal relations. As stressed by Wearing and McGehee (2013), voluntourists' motivations for being engaged are not really important; what is fundamental is the chance offered by voluntourism to create space for encounters, which can have different (positive or negative) impacts on both parties (voluntourists and host community). The quality of the outcome will be strictly dependent on the subjectivities involved. This is what motivates us to suggest the reading of this specific market niche through the lens of relational goods and under the umbrella of the humanistic management which, as stated, calls for a transition from a purely economic orientation toward a more human-centric approach, in order to alleviate human suffering and to contribute to life conduciveness and dignity.

We believe our contribution to be twofold. On one hand, we are coupling the concept of voluntourism with the theory of relational goods, adding another conceptual instrument and framework to analyze it, because until now it was only analyzed by the theory of Social Distance (Woosnam & Lee, 2011), or via the Social Exchange Theory (Paraskevaidis & Andriotis, 2017). On the other hand, by analyzing voluntourism through the lens of relational goods, we introduced a conceptual framework which may lead managers and practitioners to concentrate on those aspects connected to the idea of putting people and their wellbeing at the center (Humanistic Management), for example, developing new indicators that may take care of these aspects.

As to the limitations of the study, it must be acknowledged that the conceptual nature of this chapter will need further empirical research, which should guide further studies and experimentations on personal and communitarian impact, bearing in mind that the specificity of the context or experiences selected to be analyzed can always represent a limit in reliability and repeatability (Limata, 2017).

References

Antoci, A., Sacco, P. & Vanin, P. (2007). Social capital accumulation and the evolution of social participation. *The Journal of Socio-Economics*, *36*, 128–143. doi:10.1016/j.socec.2005.11.011.

Ashley, Caroline & Roe, Dilys (2002). Making tourism work for the poor: strategies and challenges in southern Africa. *Development Southern Africa*, *19*(1), 61–82.

Bailey, A. & Russell, K. (2012). Volunteer tourism: Powerful programs or predisposed participants?. *Journal of Hospitality and Tourism Management*, *19*. doi: 10.1017/jht.2012.14.

Barnard, C. I. (1938). *The functions of the executive*. Cambridge, MA: Harvard University Press.

Becchetti, L., Pelloni, A. & Rossetti, F. (2008). *Relational goods, sociability, and happiness*. CEIS Working Paper No. 117. doi: http://dx.doi.org/10.2139/ssrn.1115838.

Becchetti, L., Degli Antoni, G. & Faillo, M. (2010b). Let's meet up! The role of relational goods in promoting cooperation. *Journal of Socio-Economics*, *39*, 661–669. doi: 10.1016/j.socec.2010.06.011.

Becchetti, L., Degli Antoni, G. & Faillo, M. (2010a). *Social distance, cooperation and other regarding preferences: A new approach based on the theory of relational goods.* Econometica Working Paper No. 18.

Becchetti, L., Trovato, G. & Londono-Bedoya, D. (2011). Income, relational goods and happiness. *Applied Economics*, *43*, 273–290. doi: 10.1080/00036840802570439.

Benson, A. M. (2011). Volunteer tourism: Theory framework to practical applications. *Contemporary geographies of leisure, tourism and mobility*. London: Routledge.

Benson, A. M. & Henderson, S. (2011). A strategic analysis of volunteer tourism organisations. *The Service Industries Journal*, *31*(3), 405–424. doi: 10.1080/02642060902822091.

Brocke, Jan vom, et al. (2009). "*Reconstructing the giant: On the importance of rigour in documenting the literature search process.*" https://aisel.aisnet.org/cgi/viewcontent.cgi?article=1145&context=ecis2009.

Brown, S. (2005). Travelling with a purpose: Understanding the motives and benefits of volunteer vacationers. *Current Issues in Tourism*, *8*, 479–496.

Bruni, L. (2008). *Per una economia capace di felicità Spunti di riflessione e provocazioni teoriche*. Available at: https://www.researchgate.net/publication/266597277_Per_una_economia_capace_di_felicita_Spunti_di_riflessione_e_provocazioni_teoriche.

Bruni, L. (2009). *L'impresa civile: Una via italiana all'economia di mercato*. Milano: EGEA.

Bruni, L. (2010). The happiness of sociality. Economics and eudaimonia: A necessary encounter. *Rationality and Society*, 22(4), 383–406. doi: 10.1177/1043463110374500.

Bruni, L. (2011). *Felicità e beni relazionali*. Retrieved from: http://www.journaldumauss.net/IMG/pdf/FELICITa-beni_rel.pdf.

Bruni, L. (2013). *Relational goods: A new tool for an old issue*. Estudos Contemporâneos da Subjetividade – ECOS, Número Temático sobre Psicologia Positiva e Felicidade Publica.

Bruni, L. (2014). Lezioni di economia civile. *Dottorato in Scienze Dell'Economia Civile, Istituzionale, Storia, Diritto Nella Società Globale*. LUMSA.

Bruni, L. & Zamagni, S. (2004). Economia civile. Efficienza, equità, felicità pubblica. Il mulino.

Bruni, L. & Sugden, R. (2006). Fraternity: A moral understanding of market relationships. Available at: https://www.researchgate.net/publication/238660100.

Bruni, L. & Zarri, L. (2007). *La grande illusione - False relazioni e felicità nelle economie di mercato contemporanee*. Aiccon Working Paper No. 37.

Bruni, L. & Stanca, L. (2008). *Watching alone: Relational goods, television and happiness*. Working Paper No. 90, University of Milano-Bicocca, Department of Economics.

Butcher, J. & Smith, P. (2010). Making a difference. *Volunteer Tourism and Development, Tourism Recreation Research*, *35*(1), 27–36. doi: 10.1080/02508281.2010.11081616.

Coghlan, A. (2015). Prosocial behaviour in volunteer tourism. *Annals of Tourism Research*, *55*, 46–60. doi: 10.1016/j.annals.2015.08.002.

Cohen, E. (1984). The sociology of tourism: approaches, issues, and findings. *Annual Review of Sociology*, *10*(1), 373–392.

Conran, M. (2011). They really love me! Intimacy in Volunteer Tourism. *Annals of Tourism Research*, *38*(4), 1454–1473. doi: 10.1016/j.annals.2011.03.014.

Crouch, David (2002). *Surrounded by place: embodied encounters. Tourism: Between place and performance*, 207–218.

Donati, P. (1986). *Introduzione alla sociologia relazionale*. Milan: Franco Angeli.

Donati, P. (2003). Giving and Social Relations: The culture of free giving and its differentation today, *International Review of Sociology*, *13*(2), 243–272.

Donati, P. (2005). La sociologia relazionale: una prospettiva sulla distinzione umano/non umano nellescienze sociali. *Nuova umanità*, *157*, p. 97–122.

Ellis, C. (2003). Participatory environmental research in tourism: A global view. *Tourism Recreation Research*, *28*(3), 45–55.

Fee, L. & Mdee, A. (2011). How does it make a difference? Towards accreditation of the development impact of volunteer tourism. In A. M. Benson (Ed.), *Volunteer tourism: Theory framework to practical applications* (Contemporary Geographies of Leisure, Tourism and Mobility). London: Routledge.

Fink, A. (2005). *Conducting research literature reviews: From the Internet to paper* (2nd ed.). Thousand Oaks, CA: Sage Publications.

Germann Molz, J. (2015). Giving back, doing good, feeling global: The affective flows of family voluntourism. *Journal of Contemporary Ethnography*, 46. doi: 10.1177/0891241615610382.

Giddens, A. (1994). Living in a post-traditional society. In Ulrich Beck, Anthony Giddens & Scott Lash (Eds.), *Reflexive modernization: politics, tradition and aesthetics in the modern social order* (pp. 56–109). Cambridge: Polity Press.

Gray, N. J. & Campbell, L. M. (2007). A decommodified experience? Exploring aesthetic, economic and ethical values for volunteer ecotourism in Costa Rica. *Journal of Sustainable Tourism*, *15*(5), 463–482.

Gui, B. (1987). Eléments pour une définition d'économie communautaire. *Notes et documents*, *19–20*, 32–42.

Gui, B. (2000). Beyond transactions: On the interpersonal dimension of economic reality. *Annals of Public and Cooperative Economics*, *71*(2), 139–169.

Hall, C. M., ed. (2007). *Pro-poor tourism: who benefits?: Perspectives on tourism and poverty reduction*. Vol. *3*. Channel View Publications.

Henderson, K. A. & Presley, J. (2003). Globalization and the values of volunteering as leisure. *World Leisure Journal*, *45*(2), 33–37.

Kant, I. (1997). *Critique of practical reason*. Cambridge, UK; New York: Cambridge University Press.

Lee, Y. J. & Woosnam, K. (2010). Voluntourist transformation and the theory of integrative cross-cultural adaptation. *Annals of Tourism Research*, *37*, 1186–1189. doi: 10.1016/j.annals.2010.04.012.

Limata, P. (2017). Assessing social impact through theory of change in voluntourism. 20th Excellence in Services International Conference, Conference Proceedings. Available at: http://sites.les.univr.it/eisic/wp-content/uploads/2018/07/20-EISIC-Limata.pdf.

Loiseau, B., Sibbad, R., Raman, S. A., Darren, B., Loh, L. C. & Dimaras, H. (2016). Perception of the role of short term volunteerism in international development: Views from volunteers, local hosts and community members. *Journal of Tropical Medicine*. doi: https://doi.org/10.1155/2016/2569732.

Loza, A., Cristian, R. & Habisch, A. (2013). Humanistic management. *Handbook on the economics of reciprocity and social enterprise*. https://www.elgaronline.com/abstract/9781849804639.00026.xml.

Lupoli, A. C. & Morse, W. (2014). Assessing the local impacts of volunteer tourism: Comparing two unique approaches to indicator development. *Social Indicators Research*, *120*. doi: 10.1007/s11205-014-0606-x.

Lyons, K. D. & Wearing, S. (Eds.) (2008). *Journeys of discovery in volunteer tourism – international case study perspective*. Wallingford, UK: CABI Publishing.

Lyons, K., Hanley, J., Wearing, S. & Neil, J. (2012). Gap year volunteer tourism: Myths of global citizenship?. *Annals of Tourism Research*, *39*(1), 361–378. doi: 10.1016/j.annals.2011.04.016.

McGehee, N. G. & Santos, C. A. (2005). Social change, discourse and volunteer tourism. *Annals of Tourism Research*, *32*(3), 760–779.

McMillon, B., Cutchins, D. & Geissinger, A. (2006). *Volunteer vacations* (9th ed.). Chicago, IL: Chicago Review Press.

Mdee, A. & Emmott, R. (2008). Social enterprise with international impact: The case for fair trade certification of volunteer tourism. *Education, Knowledge and Economy*, 2(3), 191–201.

Meadows, D. H. (1997). *Places to intervene in a system*. Whole Earth Winter.

Melé, D. (2003). The challenge of humanistic management. *Journal of Business Ethics*, *44*(1), 77–88.

Melé, D. (2011). The firm as a "community of persons": A pillar of humanistic business ethos. *Journal of Business Ethics*. doi: http://dx.doi.org/10.1007/s10551-011-1051-2 (accessed 15 October 2011).

Mion, G. & Adaui, C. L. (2011). *Verso il metaprofit: Gratuità e profitto nella gestione d'impresa*. Siena: Cantagalli.

Mustonen, P. (2005). *Volunteer tourism – Postmodern pilgrimage?*. National, European, Global. Research Seminar of Economic Sociology – Series Discussion and Working Paper No. 2.

Nigri, G. & Michelini, L. (2019). A systematic literature review on Social Impact Assessment: Outlining main dimensions and future research lines. *International dimensions of sustainable management: Latest perspectives from corporate governance, responsible finance and CSR*. Switzerland: Springer.

Novelli, M., ed. (2005). *Niche tourism: Contemporary issues, trends and cases*. Routledge.

Nussbaum, Martha C. (2011). *Creating capabilities*. Harvard University Press, 2011.

Palacios, C. (2010). Volunteer tourism, development and education in a post-colonial world: Conceiving global connections beyond aid. *Journal of Sustainable Tourism*, *18*(7), 861–878.

Paraskevaidis, P. & Andriotis, K. (2017). Altruism in tourism: Social Exchange Theory vs Altruistic Surplus Phenomenon in host volunteering. *Annals of Tourism Research*, *62*, 26–37. doi: 10.1016/j.annals.2016.11.002.

Park, R. E. (1924). The concept of social distance as applied to the study of racial attitudes and racial relations. *Journal of Applied Sociology*, *8*, 339–344.

Pompurová, K., Marčeková, R., Šebová, L., Sokolová, J. & Žofaj, M. (2018). Volunteer tourism as a sustainable form of tourism—The case of organized events. *Sustainability, MDPI, Open Access Journal*, *10*(5), 1–12.

Poon, A. (1993). *Tourism, technology and competitive strategies*. CAB international.

Raymond, E. (2008). 'Make a difference!': The role of sending organizations in volunteer tourism. In K. Lyon & S. Wearing (Eds.), *Journeys of Discovery in Volunteer Tourism* (pp. 49–60). Cambridge, MA: CABI Publishing.

Raymond, E. M. & Hall, M. H. (2008). The development of cross cultural (mis) understanding through volunteer tourism. *Journal of Sustainable Tourism*, *16*(5).

Schneller, A. J. & Coburn, S. (2018). For-profit environmental voluntourism in Costa Rica: Teen volunteer, host community, and environmental outcomes. *Journal of Sustainable Tourism*, *26*(5), 832–851. doi: 10.1080/09669582.2017.1421200.

Shepherd, R. (2002). Commodification, culture and tourism. *Tourist Studies*, *2*(2), 183–201.

Simpson, K. (2004). 'Doing development': The gap year, volunteer-tourists and a popular practice of development. *Journal of International Development*, *16*, 681–692. doi: 10.1002/jid.1120.

Sin, H. L. (2009). Volunteer tourism - "Involve me and I will learn?". *Annals of Tourism Research*, *36*, 480–501. doi: 10.1016/j.annals.2009.03.001.

Sin, H. L. (2010), Who are responsible to? Locals tales of volunteer tourism. *Geoforum*, *41*(6), 983–992.

Singh, D. R. (2014). Volunteer tourism and host community. *International Journal of Scientific Research and Management*, 2(10). Retrieved from: https://ijsrm.in/index.php/ijsrm/article/view/1120.

Skoczylas, L. (2012) Volunteer tourism from the perspective of the social sciences. *Tourism*, *22*(2). doi: 10.2478/v10106-012-0009-9.

Smith, A. (1894). *A Catalogue of the Library of Adam Smith: Author of the 'Moral Sentiments' and 'The Wealth of Nations'*. Macmillan and Company.

Spitzeck, H., Pirson, M., Amann, W., Khan, S. & von Kimakowitz, E. (2009). *Humanism in Business*. Cambridge: Cambridge University Press.

Stainton, H. (2016). A segmented volunteer tourism industry. *Annals of Tourism Research*, 61. doi: 10.1016/j.annals.2016.09.011.

Suanpang, P., Srisuksai, N. & Tansutichon, P. (2018). The demand of voluntourism in a developing country. *Journal of Service Science and Management*, *11*, 333–342. doi: 10.4236/jssm.2018.113023.

The Guardian. (2007). *You're better off backpacking. The Guardian*. Article published on August 14, 2007. Available at: https://www.theguardian.com/uk/2007/aug/14/students.charitablegiving.

Tomazos, K. (2016). *Volunteer tourism and living with contradiction: Symbiosis or just compromise?*. In ATLAS Annual Conference, Canterbury, United Kingdom, 115–118.

Tomazos, K. & Cooper, W. (2011). Volunteer tourism: At the crossroads of commercialisation and service? *Current Issues in Tourism*. doi:10.1080/13683500.2011.605112.

Tomazos, K. & Butler, R. (2012). Volunteer tourists in the field: A question of balance?. *Tourism Management*, *33*, 177–187. doi: 10.1016/j.tourman.2011.02.020.

Tourism Research and Marketing (TRAM) (2008). *Volunteer tourism: A global analysis*. Arnhem: ATLAS Publications.

Tranfield, D., Denyer, D. & Smart, P. (2003). Towards a methodology for developing evidence-informed management knowledge by means of systematic review. *British Journal of Management*, *14*, 207–222.

Tsao, F. C. & Laszlo, C. (2019). *Quantum leadership: New consciousness in business*. Stanford, CA: Stanford University Press.

Uhlaner, C. (1989). Relational goods and participation: Incorporating sociality into a theory of rational action. *Public Choice*, *62*, 253–285.

Ulrich, P. (2009). Towards a civilized market economy: Economic citizenship rights and responsibilities in service of a humane society. In H. Spitzeck, M. Pirson, W. Amann, S. Khan & E. von Kimakowitz (Eds.), *Humanism in Business* (pp. 143–155). Cambridge: Cambridge University Press.

Vodopivec, B. & Jaffe, R. (2011). Save the world in a week: Volunteer tourism, development and difference. European Journal of Development Research, 23. doi: 10.1057/ejdr.2010.55.

VolunTourism International (1999). *VolunTourism*. Avialable at: http://www.voluntourism.org/index.html.

Wearing, S. (2011). *Volunteer tourism – Experiences that make a difference*. Wallingford, UK: CABI Publishing.

Wearing, S. & McGehee, N. G. (2013). Volunteer tourism: A review. *Tourism Management*, *38*(C), 120–130.

Webster, J. & Watson, R. T. (2002). *Analyzing the past to prepare for the future: Writing a literature review*. MIS quarterly, xiii–xxiii. https://www.jstor.org/stable/4132319.

Wong, J., Newton, J. & Newton, F. (2014). Effects of power and individual-level cultural orientation on preferences for volunteer tourism. *Tourism Management*, *42*, 132–140. doi: 10.1016/j.tourman.2013.11.004.

Woosnam, K. & Lee, J. Y. (2011). Applying social distance to voluntourism research. *Annals of Tourism Research*, *38*, 309–313. doi: 10.1016/j.annals.2010.06.003.

Young, A. F. & Young, R. (2008). Measuring the effects of film and television on tourism to screen locations: A theoretical and empirical perspective. *Journal of Travel & Tourism Marketing*, 24(2-3), 195–212.

Zamagni, S. (2006). *L'economia come se la persona contasse: Verso una teoria economica relazionale*. Aiccon Working Paper No. 32.

4 Humanistic Tourism Through The Mayan Train Megaproject

A New Narrative for Tourism Development in the Yucatán Peninsula

Blanca A. Camargo, Mario Vázquez-Maguirre, and Alfonso Ernesto Benito

1 Introduction

The Mayan Train is one of the megaprojects proposed by the 2018-2024 left-wing government of Mexico. Through 1,525 kilometers of railway, it seeks to connect five states in Southern Mexico that host Mayan archaeological sites, protected natural areas, and Indigenous communities that will inevitably be exposed to mass tourism. The government claims that it will create development opportunities for historically marginalized groups; however, there is strong opposition from local stakeholders who believe the Mayan train will benefit only rich investors and the government and will exploit Mayan cultural and natural resources for the purposes of tourism, adversely affecting the dignity, culture, and lifestyle of the Indigenous communities. Sustainable tourism and alternative forms of tourism have attempted to counterbalance the negative effects of mass tourism, but as they tend to prioritize environmental protection and income distribution, they have yet to consider human well-being and flourishing, of both the tourists themselves and the local residents, as a specific aim of tourism, particularly in post-colonial contexts where the Indigenous population has been subjected to systematic discrimination and marginalization.

A new approach to the development and management of tourism is needed if the government wants to improve the well-being of the local residents of the Yucatán peninsula by means of the Mayan Train. Humanistic management is an approach that seeks to make human well-being the aim of any organization. As human vulnerabilities materialize in certain contexts, there is an intrinsic, unconditional, and universal value to human life that needs to be protected, and also a sense of self-worth and self-respect that needs to be promoted (Pirson, 2019). Organizations, therefore, represent ways of protecting and promoting human dignity.

There is a greater need for the recognition of human dignity in everyday life, especially among disadvantaged groups such as Indigenous communities. In this chapter, we hope to contribute to a new narrative that suggests methods of organization, cooperation, and operation that will protect and promote human dignity and attempt to answer Caton's (2014) challenging question: Is it possible to outline general principles by which to promote compassion, happiness, and human dignity through tourism? Drawing upon humanistic management literature, we pay particular attention to the potential and the responsibility of the government and private industry in recognizing the human dignity of the Mayan residents in communities around the Mayan Train stations. We start off by reviewing key approaches to tourism development, and challenges related to Indigenous tourism development. We then introduce the Maya Train project to illustrate the need for a different narrative for tourism. In the last section, we discuss which tenets of humanism and humanistic management, and how, can be applied to tourism, specifically to the Mayan Train.

2 Sustainable and Alternatives Forms of Tourism: The Humanistic Gaps

The Mayan Train project is promoted as a sustainable form of social and economic development for the South-Eastern states of Mexico. It seeks to stimulate the well-being of Mayan communities by creating jobs, economic opportunities, and local development around its projected 15 stations, while protecting the ecological biodiversity in the surrounding areas (AMLO, 2018). The federal government is explicitly promoting the Mayan Train as a form of 'inclusive tourism', which respects Indigenous communities, their traditions, and their ecosystems, and where all communities will receive a fair share of the benefits derived from tourism development (Gobierno de México, 2018). This vision reflects a desire to follow the principles of sustainable tourism that started to emerge in the 1980s in response to the escalating impact of mass tourism. Although noble, sustainable and other alternative forms of tourism (e.g., ecotourism, responsible tourism, pro-poor tourism) may not be the most appropriate for the Mayan Train project given their aims, scope, and the particular characteristics of the places and people that will be affected by the megaproject.

Sustainable tourism, for example, emerged in political and academic agendas in the late 1980s as a remedy intended to decrease the environmental degradation and uncontrolled growth of tourism; its goal was to maintain tourism resources and operations over time with a view to generating economic benefits for tourism stakeholders. Sustainable tourism principles and indicators abound (e.g., Asmelash & Kumar, 2019; Choi & Sirakaya, 2006; Miller, 2001) in which sociocultural aspects are secondary to environmental and economic ones; however, humanistic concerns, such as dignity, happiness, or well-being, are almost non-existent or oriented solely towards the well-being of tourists. Moreover, sustainable tourism has received strong criticism for promoting a scientific, neo-liberal, pro-growth agenda (Higgins-Desbiolles, 2018; Mowforth & Munt, 2015) that mostly benefits governments and the private sector, without addressing equity and justice for local, disadvantaged, or minority populations (Jamal & Camargo, 2014).

At the micro-level, initiatives like responsible tourism have called for concrete stakeholder actions to make tourism more economically, environmentally, and socioculturally sustainable; the goal was to 'make *better places* for people to live in and better places for people to visit' [emphasis added] (International Conference on Responsible Tourism in Destinations, 2002). The priority is the enhancement of the destination while promoting other principles such as the participation of local people in decisions affecting their lives, and creating more enjoyable and educational experiences for tourists by giving them the opportunity to interact with local people and their heritage. Pro-poor tourism made progress in placing people, specifically poor people, at the center of the tourism agenda in order to improve their livelihoods through employment, economic links, and entrepreneurial opportunities; the focus was to provide net economic benefits for the poor at all levels of operation (Ashley, Goodwin, & Boyd, 2000).

More recently, Scheyvens and Biddulph (2017) introduced inclusive tourism, a type of tourism based on the notion of inclusive development, 'in which marginalized groups are engaged in ethical production or consumption of tourism and the sharing of its benefits' (p. 4). The distinctive element is the inclusion of marginalized populations in key areas of the tourism equation: production, consumption, decision-making, self-representation, and interpersonal relationships in order to address inequalities, challenge stereotypes, and encourage people to learn and understand the situation of minorities. Promoting the active participation of vulnerable groups in tourism (see also Jamal & Camargo, 2014) may be a first step in promoting their dignity, but what is missing is a set of principles that guide tourism towards enhancing the value and worth of the human beings directly and indirectly affected by it.

Furthermore, the approaches discussed above have been shy in addressing the particular circumstances that affect Indigenous groups in relation to tourism. For example, postcolonial patterns of oppression, exploitation, discrimination, and acculturation are often perpetuated by the tourism industry through labor exploitation via menial work, economic and political marginalization, commodification of their ecocultural resources, and problematic representations in tourism marketing and promotion (Camargo, 2011). Indigenous tourism, defined as "a form of (generally) rural tourism that allows the tourist to visit indigenous settlements and experience the daily lives of their members (Kunasekaran, Gill, Ramachandran, Shuib,

Baum, & Herman Mohammad Afandi, 2017), sought to generate economic benefits for Indigenous communities, minimize the negative effects of tourism in their culture and environment, and empower the Indigenous to make decisions affecting their development and interpret their own heritage for tourists (see Butler & Hinch, 2007; Carr, Ruhanen, Whitford and Lane, 2018 or Johnston, 2006, for a review of Indigenous people and tourism). However, The Maya Train is not conceived as an Indigenous tourism enterprise; rather, it is a government-sponsored mass tourism project on Indigenous land. The lack of results of sustainable tourism and other approaches to tourism in protecting Indigenous people justify the exploration of humanism and humanistic management in the context of tourism.

3 Humanism in Tourism Studies

Humanistic discourses are not new in tourism. In 1996, the Bureau International du Tourisme Social (BITS), through the Montreal Declaration, called for a humanist and social vision of tourism: the prime objective of tourism activities and initiatives, they argued, should be the full realization of each person's individual potential (BITS, 1996). From this perspective, tourism is seen as a vehicle for personal enrichment, and for cross-cultural and cross-generational understanding, as well as a means of promoting economic growth, especially among families, young people, and older groups. Recently, there has been a surge in scholarly interest in humanist approaches to tourism. For Caton (2014), a humanistic approach moves away from seeing tourism as an instrumental economic activity that thwarts human happiness and well-being and instead views it as an activity that serves the interests of social justice, equality, and human flourishing. In a later work (2016), she advocates a humanist paradigm for tourism studies in which the core value is to 'promote the flourishing and decrease the suffering of humans and other beings' (p. 51) and in which people question their own discourses so things can be done differently. She challenges readers to reflect on the possibility of outlining principles by which to promote compassion, happiness, and human dignity through tourism; but the challenge remains largely unattended, other than by a handful of scholars who have started to discuss dignity in tourism employment (See Cockburn-Wootten, 2012; Winchenbach, Hanna & Miller, 2019).

4 Humanistic Management

Humanistic management can shed light on how to promote a more humane approach to tourism. This approach seeks a change in the way business is carried out, which goes beyond economic goals to focus on upholding the unconditional human dignity of every individual (Spitzeck, 2011). The rationale for humanistic management is that economist management approaches have left a void in the way human dignity is protected, human flourishing promoted, and ethical development encouraged, and have been inadequate for solving issues related to social justice and social welfare (Pirson, 2019). Humanistic management starts from the idea that human beings are key to any organization (Pirson 2019), and these entities must also recognize, respect, and value their individual qualities (Bédard, 2003).

There is no single focus of or approach to humanistic management (Melé, 2016). Most authors relate the construct to concepts such as human dignity, people's well-being, ethical behavior, respect, human virtue, and dialogue between participants (Arandia & Portales, 2015; Dierksmeier, 2016; Melé, 2003, 2013, 2016; Pirson, 2019). However, every approach seems to follow a similar direction: to foster unconditional human dignity (Spitzeck, 2011) by serving the societies in which the organizations operate and increasing their citizens' well-being (Melé, 2003). In this sense, Pirson (2019) states that human dignity is indeed a necessary condition for well-being. Melé (2016) suggests seven propositions that are implied by humanism, and consequently by humanistic management, from which we posit that humanist tourism must take into consideration:

1 *Wholeness*: That is, it is reductive to view a person as merely a few aspects of their character, or according to the activity they perform or a role they play. The human being is complex and embodies myriad motivations, emotions, personality, character, capacities, spirituality, etc.
2 *Comprehensive knowledge*: Humanism seeks to understand human beings in a holistic and integral way in their different dimensions (spiritual, rational, emotional).
3 *Dignity*: The dignity of individuals is an intrinsic, absolute, and non-transferable element of the human being and, therefore, humanism aims to defend and propagate it in the entire human race.
4 *Development:* The human being is an individual in continuous growth and development. Humanism attempts to provide the necessary conditions and resources for people to grow and flourish in an integral and holistic way.
5 *Common good:* Humanism looks to benefit and maintain a harmonious balance between individual freedoms and common welfare, the outcome of people's social being. Consequently, humanism is committed to respecting free initiative, dialogue, participation, cooperation, and oneness in social life.
6 *Stewardship–sustainability:* Humanism recognizes the interconnection of the human being with the environment and all living things and promotes a sense of stewardship.
7 *Transcendence:* Humanism considers the human being as a spiritual and transcendent entity, who seeks meaning to the things they do in life and is respectful of others' religions and beliefs.

Pirson and Steckler (2018) propose a model of four independent innate drives that, when balanced, ensure basic human dignity and increase well-being. The first two – the drive to acquire (to get what we need to stay alive and produce progeny) and the drive to defend (to protect what we have acquired) – are biological and can explain the economist perspective on human nature. The other two – the drive to bond (to build long-term, mutually caring relationships with other humans) and the drive to comprehend (to make sense of the world around us; to understand ourselves and our environment; to have a basic purpose in life) – represent a humanistic extension that needs to be balanced along with the first two drives. The authors argue that these four drives represent the ultimate motives that underlie all decisions humans make to ensure that they survive and thrive. Moreover, every drive depends on moral rules (rules that also seem to be present across cultures) that provide guidance for responsible management and leadership in areas such as seeking fair and honest exchanges, helping to enhance rather than steal or destroy another's property, sharing useful information and respecting the other's beliefs even when disagreeing, helping to protect rather than harm or abandon the other (Pirson & Steckler, 2018).

How humanistic management is implemented in organizations seems to be one of the most challenging questions in this field. The Humanistic Management Center (n.d.) suggests a three-step approach to humanistic management in organizations: (1) unconditional respect for human dignity; (2) integration of ethical reflections in management decisions; and (3) active and ongoing engagement with stakeholders. Therefore, the ultimate decision-making criterion is the legitimacy given to the decision by every stakeholder (Spitzeck, 2011). This process increases the possibility that those decisions will have a positive effect on every individual involved (Ulrich, 2008). Conversely, the outcomes of some cooperation schemes that seek mutual benefits at the expense of the rest of the society are illegal, unacceptable, and illegitimate (Spitzeck, 2011).

In this sense, humanistic management approaches and principles can help tourism studies to address issues related to equity, justice, and dignity violations for Indigenous populations. The Maya Train Megaproject case example illustrates the need for a new narrative that promotes a more humane approach to tourism; one that better protects human dignity and generates well-being for every tourism stakeholder.

5 Methodology

This chapter follows a case study research strategy based on secondary data from multiple sources. Case studies are central in social sciences, including tourism research (Beeton, 2005), and instrumental especially in areas with less accessible, unique, or complex phenomena such as tourism products with many stakeholders involved (Tasci, Wei, and Milman, in press).

It usually includes a single case that is analyzed to discover relevant variables and their relationships from multiple perspectives or sources of information (Eisenhardt, 1989; Yin, 2003). Therefore, this strategy is ideal for generating novel theory (Eisenhardt, 1989).

The Mayan Train megaproject in the south of Mexico was selected as a case example based on three elements: (1) its visibility, since the Mayan Train is one of the megaprojects of the new Mexican government; (2) its relevance, as it seeks to improve economic and social indicators by increasing tourism in communities that have traditionally been excluded from the industry; and (3) its potential impact, because the project will affect the different Mayan communities and their eco-cultural resources.

6 The Mayan Train Megaproject

The Mayan Train may be the 2018-2024 left-oriented Mexican president's most important infrastructure project. Scheduled for completion in 2023, it consists of a route spanning 1,525 kilometers with 15 stations distributed through five of the country's southern states that form part of the Mayan jungle: Chiapas, Tabasco, Campeche, Quintana Roo, and Yucatán (Gobierno de México, 2018). Its construction will require an investment of between US$6,000 million and US$8,000 million, 90% of which will be financed by private investment under the public-private partnership scheme. Tourism companies are expected to develop facilities and services in the areas around each train station, which will mean that Indigenous communities will have to grant access to *ejidos* (areas of communal land) and collaborate with the developers (Tourliere, 2019). The government expects the train to attract three million people in its first years of operation, which, it is hoped, will redistribute tourism – currently concentrated primarily in Cancún and the Riviera Maya – throughout the region (Varillas, 2019).

6.1 The Mayan Train Context

The Mayan Train will cover five states in South-East Mexico, in which 10% of the Mexican population is concentrated and which represent 9.3% of the country's total GDP (INEGI, 2015). The area is one of the most culturally diverse yet historically marginalized regions of Mexico (Table 4.1): the average poverty level for the five states reached 48.5% and extreme poverty 11.4% in 2017. Tourism is the most important economic activity in the region although Tabasco and Campeche have high reserves of oil that are exploited by PEMEX, a state-owned monopoly (CEFP, 2018). Quintana Roo is the state with the highest dependence on tourism; almost one out of every four companies is engaged in tourism, employing 44.7% of the workers in the state (SECTUR, 2016). Communities in the Yucatán Peninsula are mostly Indigenous and have a fluctuating income that is based on different activities throughout the year, which are mainly agricultural, although they also involve the raising of livestock (pigs and poultry), fishing, elaboration of handcrafts (natural fiber products, basketry, manufacture of ceramic goods, manufacture of cotton garments, weaving of hammocks), and other activities that are being promoted by NGOs and social enterprises (Comisión Nacional para el Desarrollo de los Pueblos Indígenas, 2018). Specimens of flora and fauna that constitute an important sustainable basis for alternative income include the breadnut tree, chewing gum, cocoa, parlor palm, and apiculture (Programa Selva Maya, 2019). Some inhabitants of these communities migrate seasonally to be employed as temporary workers in tourist sites along the Riviera Maya (Comisión Nacional para el Desarrollo de los Pueblos Indígenas, 2018).

The Mayan ethnic group largely inhabits the jungle region that will be affected by the project. There are 1.6 million Mayans, of whom almost 10% are illiterate (Comisión Nacional para el Desarrollo de los Pueblos Indígenas, 2018). The Mayans have polytheistic beliefs; their universe is composed of three elements or regions: the sky, the earth, and the underworld, which is inhabited by different, generally anthropomorphic, supernatural beings, who can negatively or positively influence the balance of the world, and are thus are identified as the guardians of nature. Elements of Christianity – especially saints, who are considered benefactors of the community and sanctioners of those who abuse nature – have become intermingled with traditional beliefs (Comisión Nacional para el Desarrollo de los Pueblos Indígenas, 2018). Mayans are not the only ethnic group in the region; other groups that inhabit the region are listed in Table 4.2. The cultural heritage of this region is represented by more than 6,000 archaeological sites. These vestiges of Mayan culture, combined with tropical rainforests and jungle ecosystems, form attractive destinations for mass tourism.

In relation to the natural environment, the train will pass through the Selva Maya (Mayan Forest) and several protected areas, including the UN-designated Calakmul Biosphere Reserve, the second-largest expanse of tropical forest in the Americas, after the Amazon. Covering 42,300 km^2, the Selva Maya is the largest tropical rainforest in North and Central America and the second largest in the continent as a whole. It hosts an extraordinary range of biological diversity with twenty different ecosystems and a large number of endemic species (Comisión Nacional para el Conocimiento y Uso de la Biodiversidad, 2017). The Calakmul Biosphere Reserve, which spans an area of 7,232 km^2, is recognized as a Mixed World Heritage Site by UNESCO. This reserve, together with the Balam Kú and Balam Kín reserves, is part of one of the most valuable biodiverse areas in the world, with 358 species of birds, 75 species of reptiles, 18 species of amphibians, 31 species of fish, 380 species of butterflies, and 86 species of mammals. It also has one of the largest jaguar populations in Central America (Programa Selva Maya, 2019). The government claims

Table 4.1 Cultural, environmental, and social data by state

State	*Indigenous groups*	*Archaeological Sites*	*Protected areas*	*UNESCO World Heritage sites*	*Intangible cultural heritage manifestations*	*Total population (millions)*	*Poverty (%)*	*Extreme Poverty (%)*
Quintana Roo	1	17	12	1	11	1.5	28.8	4.2
Yucatán	1	16	4	2	13	2.1	41.9	6.1
Chiapas	12	10	9	1	36	5.2	77.1	28.1
Campeche	4	17	5	2	22	0.9	43.6	6.7
Tabasco	1	5	2	0	9	2.4	50.9	11.8

Source: Compiled by the authors with data from Consejo Nacional de Áreas Naturales Protegidas (n.d.), Sistema de información cultural (n.d.), and CONEVAL (2018).

Table 4.2 Ethnic groups in the region

Ethnic Group	*Population*
Maya	1,646,782
Tseltal	689,797
Tsotsil	614,105
Ch'ol	328,867
Zoque	104,321
Chontal	57,792
Mam	27,050
Q'anjob'al	13,737
Akateko	3,202
Q'eqchi	2,138
K'iche	1,388
Jakalteco	1,315
Lacandón	1,166
Ixil	335
Kaqchikel	207

Source: Compiled by the authors with data from Comisión Nacional para el Desarrollo de los Pueblos Indígenas (2018).

the environmental impact will be minimal, as 90% of the infrastructure it needs already exists and adequate measures will be taken to protect the environment, including the development of animal corridors and reforestation areas (Gobierno de México, 2018).

6.2 Controversies Surrounding the Mayan Train

The Mayan Train has met with strong opposition from environmental and civil society groups, activists, researchers, and the private sector. They are concerned that legislation put into place to protect development projects in Indigenous communities will be disregarded, and that the government's overly optimistic view of the project stems from a lack of environmental and sociocultural impact assessments.

With regard to Indigenous issues, different laws and international agreements recognize the right of Indigenous communities to self-determination, which includes deciding their own development strategies. Article 2 of the Political Constitution of Mexico guarantees Indigenous peoples the right to self-determination and the autonomy to decide their internal forms of coexistence and social, economic, political, and cultural organization. In addition, Article 6 of the National Institute of Indigenous Peoples' Law (Ley del Instituto Nacional de los Pueblos Indígenas, 2018) guarantees the right to consultation and free and informed consent to Indigenous and Afro-Mexican peoples and communities whenever the federal government promotes legal reforms and administrative acts likely to affect them. Internationally, Article 32 of the

United Nations Declaration on the Rights of Indigenous Peoples (2007) indicates that Indigenous people have the right to determine and develop priorities and strategies for the development or use of their territories and other resources. It also mandates that the states consult and cooperate in good faith with the Indigenous communities through their own governance structure in order to obtain their free and informed consent before approving any project that has an impact upon them. In this scenario, the future of the project is likely to depend on the Indigenous people themselves, who, under Mexican law, must give their consent before a project of this scale can be built on their territory (Morgan, 2019). However, the government has not undertaken meaningful consultation nor sought informed consent with the Indigenous groups that inhabit the five states affected by the Mayan train. The only consultation activity was a nationwide survey on the most important projects and social programs planned for the next six-year term of the presidency; the Mayan Train was approved by 89% of the almost one million people who voted, but that is less than 0.8% of the total population (AMLO, 2018)[1].

Although the project is labeled as sustainable tourism by the government, environmentalists are concerned about the impact of a train going through ecological reserves such as Sian Ka'an and Lake Bacalar (Fuentes, 2018), the destruction of jungles, and the loss of ecosystems and cultural elements that make local communities unique (Reportur, 2019). Furthermore, the train seems likely to incentivize agricultural activity, since it offers cheaper transportation of agricultural products to urban areas, which may create additional pressure on the forest and jungle (Fuentes, 2018). Ultimately, any scheme to move tourism from the Riviera Maya to the jungle must take into account the capacity of the jungle to support mass tourism, something that is still unknown despite the project having already been approved (Molina, 2019). From an economic perspective, representatives of chambers of commerce have also expressed their concerns about the government's ability to finance the project, or the public–private schemes that will be promoted to attract private investment (Valle, 2018).

In summary, the Train Maya megaproject poses potential threats to Indigenous communities that could potentially affect their socioeconomic lifestyle. The presence of important players in the tourism industry can generate uneven outcomes for Indigenous people that may perpetuate neocolonial dynamics and loss of their habitat. To prevent these potential outcomes and departing from elements of humanism and humanistic management, in the next section we propose a set of principles that can help tourism become more humane and go beyond economic and environmental goals to protect human dignity and promote human flourishing and wellbeing.

7 Towards Humanistic Tourism: A New Narrative for Tourism Development

Without doubt, tourism can be planned and managed in a humanistic way: that is, in a way that puts humans – in particular Indigenous and minority groups who are most vulnerable to the impact of tourism – as the primary focus of tourism, protects their dignity, and promotes their flourishing. Drawing upon Pirson and Steckler's work (2018), we propose a set of guiding principles for humanistic tourism that can help balance the four needs that protect baseline human dignity and foster flourishing and well-being. Each of these principles is explained below.

7.1 Self-determination and Self-control

Self-determination refers to the right of Indigenous and ethnic groups to freely pursue their economic, social, and cultural development (United Nations, 2007), and is particularly relevant in the context of tourism. Cultural, ethnic, and Indigenous groups must be respected in their choice to accept or reject tourism as a form of economic development, and, should they decide to participate in tourism, to acquire representation, an independent voice, and the decision-making power to shape tourism according to their values, visions of development, and socioeconomic priorities. In a humanistic tourism paradigm, the government respects Indigenous rights and laws, and makes efforts to identify people or groups in vulnerable situations in order to promote their informed, active, timely, and effective participation in tourism decision-making. In the case of the Mayan Train, for instance, Indigenous groups live in small rural communities that democratically elect their authorities, constituting a *Comisariado de Bienes Comunales* (Comisariat of Common Goods, CCG), which manages and distributes land ownership between families and regulates the use of land for productive purposes (Comisión Nacional para el Desarrollo de los Pueblos Indígenas, 2018). In this case, the CCG in each community must approve a change in the use of land that allows the Mayan Train and its stations to be built. Self-direction can be promoted by empowering Indigenous groups to co-create tourism experiences and determine, among other things, the areas and sites they wish to open to tourists, the number of travelers they are willing to host in their communities, and the cultural aspects they are willing to share with these visitors.

7.2 Decent and Meaningful Tourism Work

Several studies document that cultural minority groups, in particular women, tend to be exploited and forced into doing menial work in the tourist industry (e.g., Camargo, 2011; Redwood, 2008). In menial work, people are underpaid, overworked, and given few opportunities to develop or exercise their skills, autonomy, or creativity; they are confined to physical and monotonous tasks with little likelihood of upward mobility. Exploitative working conditions undermine people's dignity, human agency, and the self-respect needed to live a healthy life (Hodson, 2001); they are treated as nothing more than tools, a means to an end (Kant, 1998). A humanist approach to tourism labor would protect human dignity by providing decent jobs that would allow people to escape poverty and live a dignified life (e.g., formal work contracts, fair wages in relation to the local cost of living, job security, respect for workers' rights, safety, and adequate time-off so they have time to attend to their families and participate in community and cultural life, among other things), and, following the Aristotelian notion of virtue ethics, offer opportunities for people to flourish through jobs that facilitate and enhance the possibility of their living a well-rounded life (Jamal & Menzel, 2009). In the context of tourism, this means providing workers recruited to work on the Mayan Train and associated businesses with opportunities to develop and exercise skills, exert autonomy and creativity in their jobs, and enhance their prospects for promotion to meaningful, well-paid, and socially acceptable jobs (Bolton, 2007; Hodson, 2001). This can be achieved, for example, by offering entry-level tourism employees scholarships, professional development courses, on-the-job training, or management shadowing programs to assist them in obtaining managerial positions in the tourism industry.

7.3 Prioritization of Entrepreneurial Opportunities for Indigenous People

Indigenous enterprises differ from market-driven enterprises because of differences in personal values, culture, and traditions (Dana & Anderson, 2007), their level of development, and access to resources (Vázquez-Maguirre, 2012). Instead of prioritizing economic goals, Indigenous ventures tend to privilege ecological balance, solidarity, preservation of culture (Curry, Donker & Michel, 2016), sustainability, and decent work (Vázquez-Maguirre, Portales, & Velásquez 2018). As a consequence, there seems to be a natural humanistic approach to how Indigenous communities exploit entrepreneurial opportunities since they tend to place human well-being at the center of their ventures. The Mayan Train will generate many business opportunities along its route (accommodation, restaurants, tours, souvenir shops, cultural centers); however, there is a risk that these opportunities will be misused by external entities whose priority is the generation of profit, which creates a conflict with local values and needs. Indigenous communities must be given the resources and support needed to generate enterprises that promote decent work, community development, well-being, and nurturing relationships for local citizens. In this sense, the CCG can coordinate community-based enterprises and supervise new local ventures within the community. Consequently, its autonomy must be strengthened and protected so that external entities (through strategies such as lobbying) do not obtain land concessions to do business without the community's approval. The government needs to prioritize infrastructure development, capacity building, loans and subsidies, and marketing of businesses and cooperatives run by the Indigenous community in such a way that they can fairly compete with private businesses.

7.4 Recognition

Since colonial times, Indigenous groups in Mexico have suffered from discrimination, prejudice, and belittlement which has not only resulted in displacement and loss of rights, but also had severe consequences for their identity, culture, and language (Castellanos-Guerrero, 2000, 2003). Often Indigenous and cultural groups are portrayed as noble and hard-working, but uneducated and unprepared, a stereotype that has led to their marginalization in political life and exclusion from society's decision-making processes, including those concerning tourism. Their names, resources, landscapes, and images, however, are commonly used and exoticized in tourism marketing and promotional materials with no fair compensation being paid (See Camargo, 2011; Camargo, 2019). Taylor (1994) called attention to the issue of *misrecognition*, the depreciation that imposes inferior or demeaning images on others resulting in a sense of shame and an internalized belief of inferiority by imprisoning people in a false, distorted, or reduced sense of being.

Positive recognition from peers is essential for human dignity and a positive sense of self, and the Mayan Train Project has the potential to remedy past and present misrecognitions by transforming the manner in which Indigenous groups are valued and represented in tourism. Acknowledging Indigenous groups as equals in tourism planning, development, and management, and giving them formal representation and voting rights in the decision-making that affects their lives, their culture, and/or their territory is a first step to ensuring recognition and parity of participation (Fraser, 2001). Nevertheless, transformative actions are also needed to revalue the cultural identity and status of Indigenous groups and increase their self-esteem. Tourism can be a vehicle to promote the positive representation of Indigenous groups in public spheres, to foment respectful, symmetrical, and transformational relationships with tourists (avoiding situations of servility), to revalue cultural products (crafts, foods, music, and other cultural expressions) in the tourism market, and to increase the involvement of Indigenous groups in the interpretation of their own heritage.

7.5 Oneness

Western thought is dichotomous (right/wrong; West/East; human/nature, etc.) in contrast to the Indigenous world view that does not necessarily view everything in terms of binary oppositions. In this sense, the concept of oneness represents how human beings consider themselves as part of a whole. This also has consequences in how decisions are made, because nature is also part of human dignity. Oneness is also a key element in humanist management. Melé (2016) considers oneness to be an essential element in social welfare. Previous cases of Indigenous communities in Latin America engaging in tourist activities indicate that their world view relates to achieving environmental, economic, and sociocultural sustainability (Sidali, Morocho, Garrido-Pérez, 2016; Vázquez-Maguirre, Portales, & Velásquez 2018), which in turn suggests that every dimension is equally important and part of a whole. The Mayan Train project must consider that the Indigenous world view in the potentially affected communities strongly supports the unity of the human being with nature. These communities have inhabited their territories since ancient times and any change in their environment could potentially have a negative impact on the survival of the group. Because of this, they are especially sensitive to projects such as this one that may alter their relationship with nature and that could also change the economic and cultural patterns within the community. This project, then, needs serious participatory studies with an anthropological focus on the living conditions of local people and their interactions with nature and their economic and historic needs, so that communities can evaluate the potential impact on their way of life and make informed decisions as to whether to allow tourism development in their ancestral lands.

These five principles can help balance the four needs that protect baseline human dignity and foster flourishing and well-being. The first three principles – self-determination and control, decent and meaningful work, and prioritization of entrepreneurial opportunities for Indigenous people – will help Mayan communities to balance the need to acquire the basic means to ensure their survival and the need to defend their land, their resources, their culture, and their dignity. The principles of "recognition" and "unity" will contribute to fulfilling the need to bond (establish fair relationships with other stakeholders) and the need to generate a comprehensive understanding of the development dynamics they are subject to, in such a way that these communities can modify them according to their individual values and culture. Hence, Indigenous communities may be in better position to achieve environmental, economic, and sociocultural sustainability

8 Conclusion

Indigenous people are facing extreme challenges to their economic and socio-cultural survival due to the loss of their habitat, globalization forces, and neocolonial dynamics imposed by tourism. Therefore, a new paradigm is needed since sustainable tourism has not fulfill its goals. This chapter has attempted to contribute to a new narrative that protects and promotes human dignity in activities related to tourism. Drawing upon Pirson and Steckler (2018), we have proposed five guiding principles that can help balance the four needs that protect baseline human dignity and achieve flourishing and well-being. The new narrative for tourism development must include at its core the concept of human dignity, which would imply, at least in the case of Indigenous communities, that sociocultural and environmental dimensions are as important as economic dimensions in decision making. This element differentiates this narrative from sustainable and other forms of tourism which prioritize the environmental and economic dimension and see nature as a resource for tourism instead of an important component of human dignity. Future research may analyze the effects of other mega-projects on local communities and how traditional methodologies to evaluate mega-projects may be affected by a humanistic perspective and the guiding principles for humanistic tourism proposed in this chapter. Also, future works could examine the impact of humanistic tourism on the main stakeholders involved, especially the tourist.

The dignity of host communities is inalienable and the Mayan Train may provide development opportunities for marginalized communities in the Mayan jungle, and a humanist approach to tourism will ensure that these opportunities do not come at the expense of their beneficiaries' dignity and well-being.

Note

1 Recently, the government has been accused of faking consultations with Indigenous communities regarding the 'Corredor Transístmico' megaproject (which aims to be an alternative to the Panama Canal). The invitations to roundtable discussions were issued by government bodies that did not have legal status to do so, and they did not reach every Indigenous community. From the 28 Indigenous communities in Oaxaca and Veracruz that should have been consulted, only 11 participated. Moreover, these discussions lasted less than a month, which does not seem to be enough time to understand and integrate the needs, concerns, and world view of the Indigenous communities into the planned megaproject (Meza, 2019).

References

AMLO. (2018). Programas y proyectos de AMLO obtienen aprobación ciudadana superior al 90% en la Consulta Nacional Programas Prioritarios. Comunicado 034. Retrieved from https://lopezobrador.org.mx/temas/tren-maya/.

Arandia, O. & Portales, L., (2015). Fundamentos de la gestión humanista: una perspectiva filosófica. *AD-minister*, *26*, 123–147.

Ashley, C., Goodwin, H. & Boyd, C. (2000). Pro-poor tourism: putting poverty at the heart of the tourism agenda. *Natural Resources Perspectives* (51, pp. 1–6).

Retrieved 27 May 2019 from http://dlc.dlib.indiana.edu/dlc/bitstream/handle/10535/4134/51-pro-poor-tourism.pdf.

Asmelash, A. & Kumar, S. (2019). Assessing progress of tourism sustainability: Developing and validating sustainability indicators. *Tourism Management*, *71*, 67–83.

Bédard, R., (2003). Los Fundamentos del Pensamiento y las Prácticas Administrativas: El rombo y las cuatro dimensiones filosóficas. *AD-minister*, *3*. 68–88.

Beeton, S. (2005). The case study in tourism research: a multi-method case study approach. In Camargo, B.W. Ritchie, P. Burns, C. Palmer (Eds.), *Tourism research methods: integrating theory with practice* (pp. 37–48). Wallingford: CABI Publishing.

BITS (1996). The Montreal Declaration: For a Humanistic and Social Vision of Tourism. Retrieved from http://www.larosadei4venti.org/wp-content/uploads/2019/02/Dichiarazione-di-Montreal-1996Appendice-Aubagne-2006-ENG2.pdf.

Bolton, S. (Ed.). (2007). *Dimensions of Dignity at Work*. London: Routledge.

Butler, R. & Hinch, T. (Eds.). (2007). *Tourism and Indigenous Peoples: Issues and Implications*. Oxford, UK: Routledge.

Camargo, B. (2011). *Justice and fairness in tourism: A grounded theory study of cultural justice in Quintana Roo, Mexico (Doctoral dissertation)*. College Station, Texas, United States: Texas A&M University.

Camargo, B. (2019). Palenqueras in Colombia: A case of cultural injustice in tourism. In Carr, T. Jamal, *Justice and Ethics* (pp. 72–77). London: Routledge.

Carr, A., Ruhanen, L., Whitford, M. & Lane, B. (2018). *Sustainable tourism and indigenous peoples*.London: Routledge.

Caton, K., (2014). Humanism and tourism: A moral encounter. In Caton, M. Mostafanezhad & K. Hannam (eds), *Moral Encounters in Tourism* (pp. 185–198). Surrey: Ashgate.

Caton, K., (2016). A humanist paradigm for tourism studies: Envisioning a Collective Alternative to Epistemic Literalism. In Castellanos Guerrero, A. M. Munar & T. Jamal (eds), *Tourism research paradigms: Critical and emergent knowledges* (pp. 35–56). Bingley: Emerald Group.

Castellanos Guerrero, A., (2000). *Antropología y racismo en México. Desacatos*, *4*(2000), 53–79.

Castellanos Guerrero, A., (2003). Imágenes racistas en ciudades del sureste. In Choi, A. Castellanos Guerrero (ed), *Imágenes del racismo en México* (pp. 35–142). Mexico: Plaza y Valdés, pp. 35–142.

CEFP. (2018). *Evolución de la Actividad Productiva Nacional y de las Entidades Federativas 2003–2018*. Mexico. Retrieved from http://www.cefp.gob.mx/publicaciones/documento/2018/cefp0222018.pdf.

Choi, H. & Sirakaya, E., (2006). Sustainability indicators for managing community tourism. *Tourism management*, *27*(6), 1274–1289.

Cockburn-Wootten, C., (2012). Critically unpacking professionalism in hospitality: Knowledge, meaningful work and dignity. *Hospitality & Society*, *2*(2), 215–230.

Comisión Nacional para el Conocimiento y Uso de la Biodiversidad. (2017). Selva Maya: Monitoreo de Biodiversidad y Cambio Climático. Gobierno de México. Retrieved from https://www.gob.mx/conabio/prensa/selva-maya-101932.

Comisión Nacional para el Desarrollo de los Pueblos Indígenas. (2018). Atlas de los Pueblos Indígenas de México. Retrieved from http://atlas.cdi.gob.mx/.

CONEVAL. (2018). *Informe de evaluación de la política de desarrollo social 2018: resumen ejecutivo*. Retrieved from https://www.coneval.org.mx/Evaluacion/IEPSM/IEPSM/Documents/RESUMEN_EJECUTIVO_IEPDS2018.pdf.

Consejo Nacional de Áreas Naturales Protegidas. (n.d.). Sistema Nacional de Áreas Protegidas (SINAP). Retrieved from https://www.gob.mx/conanp/acciones-y-programas/sistema-nacional-de-areas-protegidas-sinap.

Curry, J., Donker, H. & Michel, P., (2016). Social entrepreneurship and indigenous people. *Journal of Co-operative Organization and Management*, *4*(2), 108–115. doi.org/10.1016/j.jcom.2016.09.002.

Dana, L. & Anderson, R., (2007). A multidisciplinary theory of entrepreneurship as a function of cultural perceptions of opportunity. In L. P. Dana, L. P. & Dierksmeier, R. Anderson (eds), *International Handbook of Research on Indigenous Entrepreneurship* (pp. 595–604). Northampton: Edward Elgar Publishing.

Dierksmeier, C., (2016). What is 'Humanistic' About Humanistic Management? *Humanistic Management Journal*, *1*, 9–32.

Eisenhardt, K., (1989). Building theories from case study research. *Academy of Management Review*, *14*(4), 532–550. https://doi.org/10.5465/amr.1989.4308385.

Fraser, N., (2001). Recognition without Ethics?, *Theory, Culture & Society*, *18*(2–3), 21–42. https://doi.org/10.1177/02632760122051760.

Fuentes, Y., (2018). Tren Maya: así es el ambicioso proyecto que propone AMLO y tiene un costo de miles de millones de dólares para México. BBC Mundo. Retrieved 12 May 2019 from: https://www.bbc.com/mundo/noticias-america-latina-45254080.

Gobierno de México. (2018). *¿Qué es el Tren Maya?* Retrieved 15 July 2019 from https://www.gob.mx/cms/uploads/attachment/file/455316/TM_Folleto_Espa_ol.pdf.

Higgins-Desbiolles, F., (2018). Sustainable tourism: Sustaining tourism or something more?. *Tourism Management Perspectives*, *25*, 157–160.

Hodson, R., (2001). *Dignity at Work*. Cambridge: Cambridge University Press.

Humanistic Management Center. (n.d.). *A Why, How and What in Humanistic Management*. Retrieved 23 June 2019 from http://humanisticmanagement.org/cms/events/past_events/downloads/Why_How_What_Humanistic_Management_Krakow_University_of_Economics.pdf.

INEGI. (2015). *Principales resultados de la Encuesta Intercensal 2015 Estados Unidos Mexicanos*. Retrieved from http://internet.contenidos.inegi.org.mx/contenidos/productos//prod_serv/contenidos/espanol/bvinegi/productos/nueva_estruc/702825078966.pdf.

International Conference on Responsible Tourism in Destinations. (2002) The Cape Town Declaration, Cape Town. Retrieved 12 June 2019 from https://responsibletourismpartnership.org/cape-town-declaration-on-responsible-tourism/.

Jamal, T. & Camargo, B., (2014). Sustainable tourism, justice and an ethic of care: Toward the just destination. *Journal of Sustainable Tourism*, *22*(1), 11–30.

Jamal, T., Camargo, B. & Wilson, E., (2013). Critical omissions and new directions for sustainable tourism: A situated macro–micro approach. *Sustainability*, *5*(11), 4594–4613.

Jamal, T. & Menzel, C., (2009). Good Actions in Tourism. In Johnston, J. Tribe (ed), *Philosophical Issues in Tourism* (pp. 227–243). Bristol: Channel View Publications.

Johnston, A. M., (2006). *Is the sacred for sale: Tourism and Indigenous peoples*. London: Earthscan.

Kant, I., (1998). *The Groundwork of the Metaphysics of Morals*. Kunasekaran, M. Gregor (ed.). Cambridge: Cambridge University Press.

Kunasekaran, P., Gill, S. S., Ramachandran, S., Shuib, A., Baum, T. & Herman Mohammad Afandi, S., (2017). Measuring sustainable indigenous tourism indicators: a case of Mah Meri ethnic group in Carey Island, Malaysia. *Sustainability*, *9*(7), 1256.

Ley del Instituto Nacional de los Pueblos Indígenas. (2018). Comisión Nacional de los Derechos Humanos. Retrieved from http://www.cndh.org.mx/sites/all/doc/Normatividad/Ley_INPI.pdf.

Melé, D., (2003). The challenge of humanistic management. *Journal of Business Ethics*, *44*(*1*), 77–88.

Melé, D., (2013). Antecedents and current situation of humanistic management. *African Journal of Business Ethics*, *7*(2), 52. https://doi.org/10.4103/1817-7417.123079.

Melé, D., (2016). Understanding Humanistic Management. *Humanistic Management Journal*, *1*(*1*), 33–55. https://doi.org/10.1007/s41463-016-0011-5.

Meza, M.A., (2019). Indigenismo simulado. Retrieved 12 May 2019 from https://elpais.com/internacional/2019/05/09/mexico/1557364435_294505.html.

Miller, G., (2001). The development of indicators for sustainable tourism: Results of a Delphi survey of tourism researchers. *Tourism management*, 22(4), 351–362.

Molina, H., (2019). Tren Maya no tendrá consecuencias ambientales. Retrieved 12 May 2019 from https://www.eleconomista.com.mx/estados/Tren-Maya-no-tendra-consecuencias-ambientales-20190514-0013.html.

Mowforth, M. & Munt, I., (2015). *Tourism and Sustainability: Development and Sew Tourism in the Third World* (3rd ed.). New York: Routledge.

Morgan, M., (2019). Mexico to 'speed up' Maya train route. Retrieved 12 May, 2019 from https://www.bbc.com/news/blogs-news-from-elsewhere-47454359.

Pirson, M., (2019). A Humanistic Perspective for Management Theory: Protecting Dignity and Promoting Well-Being. *Journal of Business Ethics*, *159*, 1–19. https://doi.org/10.1007/s10551-017-3755-4.

Pirson, M. & Steckler, E., (2018). *A Humanistic Ontology for Responsible Management Learning*. Available at: https://ssrn.com/abstract=3162851.

Programa Selva Maya. (2019). Selva Maya. Retrieved from http://selvamaya.info/es/.

Redwood, L. (2008) The Rebuilding of a Tourist Industry: Immigrant Labor Exploitation in the Post-Katrina Reconstruction of New Orleans. In Scheyvens, M. Marable & K. Clarke (eds) *Seeking Higher Ground. The Critical Black Studies Series (Institute for Research in African-American Studies Columbia University)*. New York: Palgrave Macmillan.

Reportur. (2019). Tren Maya: los indígenas critican que AMLO destruirá selvas.

Retrieved from https://www.reportur.com/hoteles/2019/04/29/tren-maya-los-indigenas-critican-amlo-destruira-selvas/.

Scheyvens, R. & Biddulph, R., (2017). Inclusive tourism development. *Tourism Geographies*, *20*(4), 589–609.

SECTUR. (2016). *Estadística Turística Derivada de los Censos Económicos 2014*. Retrieved from https://www.datatur.sectur.gob.mx/Comite%20de%20Estadisticas%20del%20Sector%20Turismo/CENSOS_ECONOMICOS_2014.pdf.

Sidali, K., Morocho, P., & Garrido-Pérez, E., (2016). Food Tourism in Indigenous Settings as a Strategy of Sustainable Development: The Case of Ilex guayusa Loes. In the Ecuadorian Amazon. *Sustainability*, *8*, 967.

Sistema de información cultural. (n.d.) Patrimonio de la humanidad. Retrieved from http://sic.gob.mx/?table=patrimonio_humanidad.

Spitzeck, H., (2011). An integrated model of humanistic management. *Journal of Business Ethics*, *99*(1), 51–62. doi: https://doi.org/10.1007/s10551-011-0748-6.

Tasci, A., Wei, W. & Milman, A. (in press). Uses and misuses of the case study method. *Annals of Tourism Research*. https://doi.org/10.1016/j.annals.2019.102815.

Taylor, C., (1994). Multiculturalism. In Tourliere, A. Gutmann (ed.), *Multiculturalism* (pp. 25–73). Princenton: Princeton University Press.

Tourliere, M., (2019). Tren Maya provocará un "choque cultural" en algunas comunidades, afirma el titular del Fonatur. Retrieved 12 May 2019 from https://www.proceso.com.mx/584103/tren-maya-provocara-un-choque-cultural-en-algunas-comunidades-afirma-el-titular-del-fonatur.

United Nations. (2007). *United Nations Declaration on the Rights of Indigenous People*. Retrieved 24 July 2019 from https://www.un.org/development/desa/indigenouspeoples/wp-content/uploads/sites/19/2018/11/UNDRIP_E_web.pdf.

Ulrich, P., (2008). *Integrative Economic Ethics: Foundations of a Civilized Market Economy*. Cambridge: Cambridge University Press.

Valle, A., (2018). Los pros y contras del Tren Maya. Retrieved 12 May, 2019 from https://expansion.mx/empresas/2018/09/28/los-pros-y-contras-del-tren-maya.

Varillas, A., (2019). Everything you need to know about the Maya Train project. Retrieved 18 May 2019 from https://www.eluniversal.com.mx/english/everything-you-need-know-about-the-mayan-train-project.

Vázquez-Maguirre, M., (2012). *Indigenous social enterprises in subsistence economies*. Monterrey: Monterrey Institute of Technology.

Vázquez-Maguirre, M., Portales, L. & Velásquez, I., (2018). Indigenous Social Enterprises as Drivers of Sustainable Development: Insights From Mexico and Peru. *Critical Sociology*, *44*(2), 323–340.

Winchenbach, A., Hanna, P. & Miller, G., (2019). Rethinking decent work: the value of dignity in tourism employment. *Journal of Sustainable Tourism*, *27*(7), 1026–1043.

Yin, R. K., (2003). *Applications of case study research*.Thousand Oaks: SAGE.

5 Ethical Value Propositions in Branding a Rural Community

Arja Lemmetyinen, Lenita Nieminen, and Johanna Aalto

1 Introduction

The focus of the research is on an evolutionary branding process (Lemmetyinen & Go, 2010) of a municipal or rural community conducted by its various stakeholders. An organization's – in this case a municipality's – internal logic, and especially its collaboration with a variety of societal stakeholders have assumed increasing importance as a mechanism for developing a reputable brand to interest not only local residents but also visitors, tourists, investors, and enterprises among others.

The research builds on identity-based place branding theory (Kavaratzis & Hatch, 2013) emerging from earlier branding theories and the contemporary marketing discourse known as the service-dominant logic (S-D logic) emphasizing the dialogue between the place branding stakeholders. In their seminal work on S-D logic, Vargo and Lusch (2004) expressed the hope that S-D logic would become a foundation for a general theory of marketing and called for the co-development of models and theories by other marketing scholars (Stanislawski, 2011). To this end, research on how S-D logic applies to branding (e.g., Beverland, Lindgren, Napoli, Ballantyne, & Aitken 2007) was undertaken, but there has not been as much focus on the ethics of S-D logic (Stanislawski, 2011; Williams & Aitken, 2011).

The authors of this study have observed the process through the lenses of both branding, and more specifically, *place branding* theories. In addition to the place branding point of view, we follow the *stakeholder concept of management* (Williams & Aitken, 2011; Freeman, 1984) as it is "integral to the theoretical base of business ethics" (Williams & Aitken, 2011, 447) recognizing "that there are more parties to business than those directly involved." Both Williams and Aitken (2011) and Stanislawski (2011) agree that S-D logic (Vargo & Lusch, 2004) is *inherently* ethical, but both researchers would like that quality to be more explicit. The authors' research experience also suggests that the ethical value proposition considerations need to be explicit for place branding in order to provide a firm ethical foundation (see Williams & Aitken, 2011, p. 448). Like Stanislawski (2011), we investigate the ethics of the stakeholder collaboration by applying Schwartz's "universal moral standards for corporate codes of ethics" as a basis from which place brand marketers "can proactively confront the ethics of co-creation".

This humanistic view is in line with the *Model of Collaborative Entrepreneurship for a More Humanistic Management* introduced by Rocha and Miles (2009, p. 446) using an Aristotelic-Thomistic approach holding that "ethics primarily deals with the true idea of human beings". Rocha and Miles state that the challenge for humanistic management is, however, to encourage leaders and managers to, in the words of Spitzeck, "talk and decide about moral issues just as they do about profitability" (Spitzeck 2011, 59). We would contend that the same challenge applies to marketing and to place branders.

In a similar way to Kabadayi et al. (2019), who devised an interdisciplinary framework utilizing the literature on both humanistic management and services, our paper applies a framework utilizing literature on S-D logic marketing and humanistic management to assist "in bridging two disciplines to highlight their potential and role in addressing the (global) challenges" (Kabadayi et al. 2019, p. 159) of introducing a humanistic view in place branding. This can be one of the first steps in a process Lovins (2016, p. 87) calls for, namely creating a new "story of a world that delivers prosperity, security and meaning; a world that works for 100% of humanity".

Our arguments follow those of Lindgreen et al. (2009) in stating that there is a relationship between marketing practice and leadership style and suggesting that those elements are dynamic and linked. Consequently, in our paper examining the ethical value propositions involved in branding a rural community, the link between the S-D logic of marketing and humanistic management is the focus on the humanistic view of each and the evident similarities in the ethical and moral values underpinning them. As Fisk (2019) notes, humanistic management and service research (and, we contend, *dominant logic marketing*) are jointly able to answer the key question regarding the role of human interaction in social innovation that is at the core of both disciplines.

Following the reasoning above we are able to define the research question of the paper as follows: How does offering ethical value propositions in a stakeholder-friendly way in place branding reflect a humanistic view of marketing and management?

The research adopts a qualitative approach (Mabey & Freeman 2012) as an informed method for examining place brand-building, viewing this process as a social system (Luhmann's theory, Luhmann 1986). The theoretical background is followed by an explanation of the methodology applied, and in the following paragraphs, we present the empirical evidence to support our preconceptions. We then draw our conclusions, before outlining the limitations of the research and potential avenues for future studies.

2 Theoretical background

2.1 Service Dominant Logic: A Stakeholder as a Value Co-creator

The concept of service-dominant logic (S-D logic) was introduced in 2004 (Vargo & Lusch 2004; Williams & Aitken 2011) and has dominated the contemporary marketing discourse ever since. It appears that S-D logic has offered the theory of marketing a more humanistic aspect than was previously available to it, by putting the customer on the same level as the firm. Moreover, as Williams and Aitken stated (2011, 441), Vargo and Lusch "claim, that S-D logic is 'inherently' ethical". The humanistic aspect is also present in an inter-organizational model fostering collaborative entrepreneurship in the human management literature (Rocha & Miles 2009).

The S-D logic approach generally advocates viewing the stakeholder as a value co-creator, who perceives the value of the experience at the very moment of the service (or the place as a *product*) in use. Therefore, the enterprise or *place marketing actors* that is the strategic network (Warnaby 2009), cannot deliver value but only *offer value propositions* to potential customers, visitors, and residents. In presenting the above-mentioned perspectives, we wish to reflect the *humanistic view of the place marketing* actors, and their ethical considerations, expressed in the place identity.

Ethics is commonly known to be something good, extremely important, or achievable. Schwartz (1992) adds that four additional dimensions of ethics are values as beliefs, as desirable behavior, transcending specific situations, and selection or evaluation of behavior. Values are considered to convey the ethics of each actor. Williams and Aitken (2011) point out that there are ~~always~~ ethical decisions involved in the voluntary component of exchange that is a feature of exchange behavior among actors. They offer an additional foundational premise to S-D logic for consideration that relates to the role of personal and societal values in the co-creation of value. The Aristotelic–Thomistic philosophical approach underpinning the collaborative entrepreneurship introduced in human management literature (Rocha & Miles 2009, 446) holds, "that ethics primarily deals with the true idea of human beings, which refers to the idea of a good man".

When collecting the value priorities of individual actors, it is possible to provide essential pieces of information about culture, even if the value propositions vary (Schwartz 2014). According to Schwartz (2014), every combination of institutions will develop a unique value culture, regardless of the existing societal value culture, by taking into consideration its "specialized tasks, challenges and membership". Linking universal moral values more to standards in business development, we follow Schwartz's (2005) theory, which is based on normative research on generalizing three sources: corporate codes of ethics, global codes of ethics, and the business ethics literature. Accordingly, six universal moral values are formulated: trustworthiness, respect, responsibility, fairness, caring, and citizenship. As a theoretical lens, we refer to Stanislawski (2011), exploring how Schwartz's (2005) six universal moral standards (trustworthiness, respect, responsibility, fairness, caring, and citizenship) could be applied to value co-creation. These moral standards are also identified in human management literature as the foundation values that should be at the core of the service processes (Kabadayi, Alkire, Broad, Livne-Tarandach, Wasieleski, & Puente 2019).

Moreover, the notion of that the consumer's *value, which is relating to a place,* will be determinable from the extent and nature of the interaction with the place (Vargo & Lusch, 2008). Warnaby (2009) has applied the seminal service-dominant (S-D) logic of Vargo and Lusch (2004) specifically to *place* marketing. Applying the S-D logic, the place consumer (visitor, tourist, or resident) can be viewed as "an individual place brand co-creator" and, extending the terms of Vargo and Lusch (2008) those responsible for the management and marketing of places offer value propositions derived from the array of facilities and resources available to them.

2.2 *Identity-based Place Branding: An Interaction and Dialogue between Stakeholders*

Following the ideology of S-D logic, Kavaratzis and Hatch (2013) introduced a novel approach to place branding theory that emphasizes the nature of place branding as an interaction and dialogue between stakeholders. This is in line with the idea of place branding not only attracting new businesses and visitors but also primarily increasing the value of the place in the eyes of the inhabitants (Moilanen & Rainisto, 2009). The individuals (in the firms and organizations) participating in the brand-building process do not have a common organizational structure, but their functions occur in the same regional area. From the customer's viewpoint, a place creates one entity, and this makes the situation more complex than the brand of a single service company (Moilanen & Rainisto, 2009).

Kavaratzis and Hatch (2013) see a link between a place brand and place identity and argue that place identity should be thought of as a complex process of identity construction rather than a specific outcome of such a process. In a similar vein, place branding is best understood as an ongoing dialogue. Therefore, the stakeholders became the most important element of the place branding process as is also pointed out in the human management literature introducing the concept of entrepreneur collaboration (Rocha & Miles, 2009).

Aaker (2014) also emphasizes how a powerful brand is built steadfastly from the inside out. During the brand-building process, the personality values of the commonplace must be identified (Datzira-Masip & Poluzzi, 2014). The successful internal communication between the actors is the core of correct and clear brand images (Moilanen & Rainisto, 2009). This coincides with the ideas of Lindgreen, Palmer, Wetzels, & Antioco (2009) emphasizing the linkages between and dynamics of human management and marketing.

Warnaby (2009, p. 415) echoing Vargo and Lusch (2008) state that "the place brand will constitute the interface between the place (and the various activities within) and the customer, and that the communication at this interface will be a two-way process". The challenge, however, is to maintain brand communication through changes among those involved in the branding process so that the brand that customers have contact with through various sources seems to be consistent (Aaker, 2004). The brand should be in a constant process of renewal and a brand portfolio audit should be conducted to analyze its development and identify options to maintain that development. As Aaker says, brand management is also a matter of organizing (Aaker, 2014). When branding a place, it is necessary to realize that every single actor (organization, entrepreneur, or representative of the municipality/city) involved in the process is related to the identity of the place in their own way, thus reflecting their identity. Structuring the place as a multi-stakeholder brand structure is one form of brand portfolio, which in the later phases of the process is formulated as brand architecture. The brand portfolio specifies the roles, relationships, and the scope of the portfolio brands (Aaker, 2004).

With regard to the process-based brand constructing approach, Kavaratzis and Hatch (2013) draw parallels with the dynamics of organizational identity (Hatch & Schultz 2002) and suggest a similar model of the dynamics of place branding. The core idea in that model (Kavaratzis & Hatch, 2013) is that effective place branding expresses the cultural understandings of the place, and at the same time taps into the images of others and mirrors their impressions and expectations. According to Kavaratzis and Hatch (2013, p. 79) "applying the Hatch and Schultz (2002) model of identity dynamics, branding can also be seen as an instrument of expressing – it expresses internal, cultural understandings of who 'we' are as a community or an organization."

2.3 *Ethical Value Propositions: A Humanistic Management View of Place Branding*

Incorporating a humanistic view on place branding is challenging because doing so demands the leaders of the process, the place branders, report moral and ethical considerations in a similar manner as they do with profitability (Spitzeck 2011). As Spitzeck (2011, p. 59) puts it while referencing Kant, if the leaders do not do so, “humanistic management will never succeed, as human dignity has a value, but not a price (Kant, 1997)”.

As Kabadayi et al. (2019) point out, the most common foundational values should be identified in the service processes and interactions involving the stakeholders in order to ensure the humanistic management approach manifests in the activities of organizations. We would contend that those service processes and interactions should be extended to include the *value propositions* made by suppliers. The interdisciplinary framework of Kabadayi et al. (2019) is praised by Fisk (2019, p. 187) who sees it as “a first step in bridging two disciplines to highlight their potential and role in addressing the global challenges”. This study seeks to follow the interdisciplinary approach of Kabadayi et al. (2019) and discuss the human management approach in the context of place branding and service-dominant logic marketing. Just as Rocha and Miles (2009) emphasized entrepreneurial collaboration in humanistic management, Kavaratzis (2012) directed attention to the emergence of entrepreneurial modes in place marketing (urban *and rural – emphasis added by the authors of this paper)* whereas Kavaratzis and Ashworth (2008) argue that a place brand should operate as an umbrella brand covering a multitude of stakeholders and audiences; only doing so one can facilitate consequence in the articulation of place brand value propositions (Warnaby 2009).

Kavaratzis and Hatch (2013) see place branding as an effective tool for allowing the local population to express the cultural features of the place. This is illustrated in Figure 5.1 below with the arrows starting from the left side of the figure. Place branding also leaves impressions on others (the place image on the right side of Figure 5.1) and the changes in the external image will be mirrored in the brand identity (Kavaratzis & Hatch 2013). Kavaratzis and Hatch (2013) base the idea of how place branding shadows the brand identity process on the earlier work of Hatch and Schultz (2002) analyzing the dynamics of organizational identity. Place branding theories have evolved toward the *brand*-identity-based *approach*, and, for example, the conceptualization of Kavaratzis and Hatch (2013) emphasizes the ongoing dialogue between all the stakeholders in the brand-building process.

According to the conceptualization of Kavaratzis and Hatch (2013) the place culture is also present in the identity of the place and is integral to its image. Prior studies (Lemmetyinen, Nieminen, and Aalto, 2018) show the prerequisites for the stakeholder collaboration. Benefits, risks, interaction, coordination, and keenness have been found to representing the dimensions of actors’ engagement in a cooperation initiative.

As Williams and Aitken (2011, p. 449) state, the ethical imperatives in

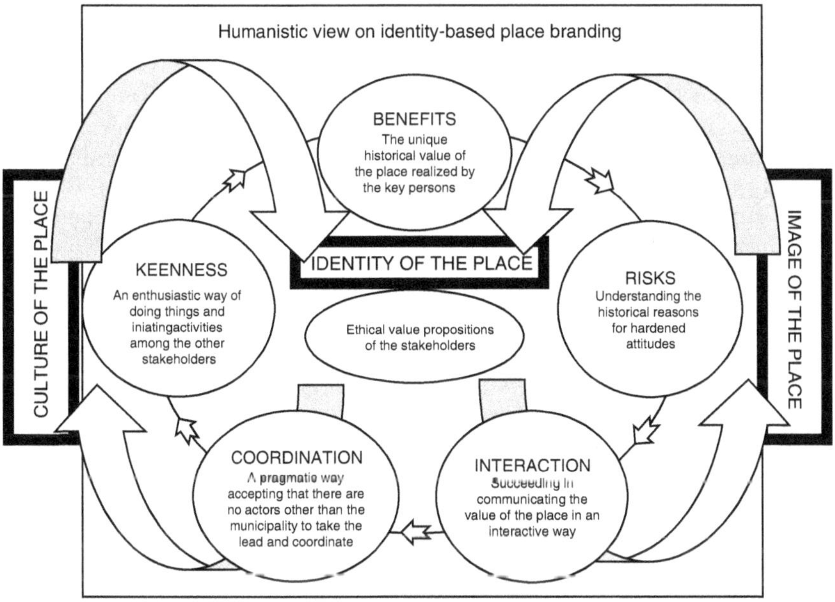

Figure 5.1 The framework of the study.
Source: Author's own elaboration based on Kavaratzis and Hatch (2013); Lemmetyinen, Nieminen and Aalto (2018).

S-D logic are implicit and should be made explicit. This reasoning gives us an option to add to Figure 5.1 a humanistic management view on the identity-based place branding thus following Williams and Aitken (2011, p. 448) statement that "ethical considerations are implicit in S-D logic and need to be more explicit for marketing (here: place branding) to have a firmer ethical foundation". In a similar vein, we have added ethical value propositions of the collaborating stakeholders in the middle of Figure 5.1 thus bringing the ethical considerations more explicitly into the stakeholder orientation.

As a synthesis from the theoretical discussion, we combined the prerequisites of *stakeholder collaboration* in branding a place (Lemmetyinen et al., 2018) and the *identity-based place branding* model of Kavaratzis and Hatch (2013) that indicates how both the culture and image of a place are expressed in its identity. The *ethical value propositions* of the stakeholders derived by adopting the *humanistic view on management and marketing* are also illustrated in Figure 5.1 above. As discussed earlier the stakeholder orientation, which is fundamental not only in the S-D logic (Williams & Aitken, 2011), but also in humanistic management (Rocha & Miles 2009), links an ethical dimension to the identity-based place branding discourse.

3 Methodology

This study focuses on the ethical value propositions of stakeholders collaborating in place branding by analyzing the values and ethics related to the place in the process of branding a rural community. To understand society as a social system, one must know not only the theory but also the *context* (Luhmann, 2012). The context of this study is a rural area in Finland to which we assigned the pseudonym The Ironworks Village. The village takes its name from an ironworks complex founded in 1689, and it has been in the hands of one of Finland's oldest family businesses since the 1870s. The place has a rich industrial history and interesting architecture, and the well-known Finnish architect Alvar Aalto left his mark there in the 1940s. The family business still owns many of the historic buildings on the site and has therefore had a major influence on the development of the place.

Our study is part of the longitudinal and intense interaction and collaboration activities undertaken during 2012–2019 with a small business community and public-sector actors of The Ironworks Village who participated in building a brand identity for their municipality. One of the authors was the personnel manager of the municipality and participated in the planning of the development projects for the area. The two other authors attended many of the related formal and informal events, for example, through student projects and assignments. By adopting an ethnographically oriented approach, the researchers were able to observe actors in the field and follow developments as they happened (Tedlock, 2000) and through those observations were able to monitor the phenomenon of stakeholder engagement in place branding over time.

We collected empirical data through in-depth interviews and formal and informal conversations (Riessman, 2004) with the municipal representatives and other public-sector agencies, private micro-business entrepreneurs, and NGOs. The themed interviews focused on the development of The Ironworks Village and the role of the informant in that development; the cooperation and competition between companies or other actors; the current and future customers of, and visitors to, the area; and also on the values of one specific business and those of The Ironworks Village. We also wanted to know what should be highlighted in the brand of The Ironworks Village and how its history and cultural heritage might affect the development of the place. The interviews were loosely structured to allow informants to address issues they wanted to discuss.

Each interview took approximately one-and-a-half hours and was tape-recorded and transcribed. The interviews used in the current study were with two micro-business entrepreneurs, the coordinator of the regional rural development organization, the head of the regional research center, the manager of the regional business agency, and the cultural and

Table 5.1 Description of the Informants

Informants	*Description*
Cultural manager of the rural municipality	Started to coordinate some of the key stakeholder groups as far back as the early 1990s when the municipality was actively involved in international co-operation through EU-funded projects; in recent years became more enthusiastic about developing cultural tourism of the area; appointed a working group to coordinate the development. "In recent years, new actors are clearly the ones who appreciate cultural heritage of the place."
Project coordinator of the regional rural development organization	Funds and runs development projects in eg fishing and nature tourism and culture in cooperation with local actors; emphasizes local democracy and collaboration
Head of the regional research center	The research center is specializing in the food economy, water protection, and restoration; funded by the regional municipalities, companies, and the local university. Would like to see the research center as part of the brand image of the place representing the core assets of the region; clean water, food production, and the cultural and natural heritage that she considers the cornerstones of the brand; runs joint research projects with the local university.
Business manager of the regional business development agency	Considers the value of the place image not only from the viewpoint of The Ironworks Village emphasizing the value of the Alvar Aalto architecture but more from that of the industrial activities of the place and the municipality and neighboring municipalities.
Personnel manager of the municipality	Has been coordinating the development of the tourism and leisure sector of the municipality for several years; was a member of the development team of the Ironworks Village at the time of the interview. Considers the cultural heritage of the area to be part of the everyday life of the people that live in the area.
Owner of the Perfume Shop	A visual and performance artist; also running a perfume shop and an art gallery. The family history and its visuality are important values for her. Continues her family's traditions by running the shop that her mother established, which was the first cosmetics retailer in the countryside in Finland. She has created service packages for tourists around her artistic activities. Active in local and regional media.

Source: Author's own elaboration

personnel managers who coordinated the brand building activity at the municipality. Table 5.1 below presents information on the interviews.

The analysis of the study is based on our sense-making and interpretation of the interviews with key actors and our observations of them (Coffey & Atkinson 1996), which combine to indicate how the culture and image of the place are expressed in its identity. All three researchers participated in the analysis and discussed their interpretations of the data. We started the analysis by reading and re-reading the transcribed interviews, and then one of the researchers organized the data manually based on the discussions around the themes assigned to fit the interests of the study, which involved color-coding excerpts from the interview data. In the following section, we present the findings arising from the analysis of our empirical data.

4 Findings and Discussion

The theoretical framework in this study utilized literature on service-dominant logic marketing and humanistic management to assist "in bridging two disciplines to highlight their potential and role in addressing the (global) challenges" (Kabadayi et al., 2019, p. 159) of introducing a humanistic view in place branding. As a synthesis from the theoretical discussion, we combined the prerequisites of *stakeholder collaboration* in branding a place (Lemmetyinen et al., 2018) and the *identity-based place branding* model of Kavaratzis and Hatch (2013) that indicates how both the culture and image of a place are expressed in its identity. The *ethical value propositions* of the stakeholders derived by adopting the *humanistic view on management and marketing* will be analysed in the following subsections as stemming from the culture of the place, the image of the place, and the identity of the place.

4.1 Ethical Value Propositions Stemming from the Culture of a Place

The informants expressed the values (Rocha & Miles 2009) they connected to the culture of the place, for example, by emphasizing the significance of the world-famous architect Alvar Aalto, whose indisputable reputation is established and respected among architects worldwide and thus draws foreign visitors from Japan and other countries. The cultural manager fully relies on the architecture when highlighting the attraction of the municipality.

"I'd put the architects in first place. People come here from Japan to see the buildings designed by Alvar Aalto. So, architecture should be a central thing in the brand of The Ironworks Village". (Cultural manager)

The coordinator of the regional rural development organization (an EU project) also emphasizes the value of the Alvar Aalto phenomenon.

She told of a Frenchman who visited the area as a member of a group on a fishing and nature trip. It turned out he was an architect who got really excited and passionate about the place and subsequently brought several groups of architects there.

"The fishing and nature tourism project has also brought a number of groups of architects here. On the first visit of the Frenchmen, by chance, there was an architect who had noticed that there were several buildings by architect Alvar Aalto in The Ironworks Village. He was a real fan of Alvar Aalto and since then he has brought several groups of French architects here. You know, it's different". (Project coordinator)

The head at the regional research center brings up the international contacts and sees how these visitors value not only the cultural heritage of The Ironworks Village but also its unique environment and the history of the area. All these citations draw a picture of *trustworthiness* behind the value proposition.

"We have had French, Chinese, Swedish and other international guests at the Lake Institute and they have all been impressed not only by this place but also the environment around here". (The head of the regional research center)

She also points out that both the nature and culture of the place should be combined in the value propositions in order to provide more business opportunities through innovative and unique offerings. This resonates with the entrepreneurial collaboration view of Kabadayi et al. (2019) as well as that of Kavaratzis (2012) focusing on the role of stakeholders and a stakeholder-oriented view of how place brands are built and also builds on the stakeholder concept of management endorsed by Williams and Aitken (2011) among others.

4.2 *Ethical Value Propositions Stemming from the Image of a Place*

A multilevel reflexive analysis of how the cultural heritage of a place could add value to its brand image helps the community to set a common vision for the brand-building. In the following citations, the informants expressed the values they connected to the image of the place (Kavaratzis & Hatch 2013). Those informants who had a form of a public position with a job description covering a wider regional area, for example, public professionals see The Ironworks Village as part of the region but do realize the risks if the brand designers are not able to respond to the challenges of potential demand.

"I'd like to emphasize that we shouldn't just market The Ironworks Village but see the brand as part of the whole region. So, it should be approached more broadly. Creating a brand is risky if we are unable to meet those challenges and demand. Of course, we should be part of the brand. We bring in the competence of water and food, so our business could be part of that brand. Those basic elements of the brand should be the diverse cultural heritage and diverse nature of this area. They are the cornerstones of the brand combined with high-quality accommodation and dining, and things like that". (Head of the regional research center)

She would like to see the regional research center as part of the brand image of the place representing the core assets of the region; clean water and food production as well as the cultural and natural heritage that she considers the cornerstones of the brand. This is a clear signal of *caring* for the wider aspects in the value propositions. It is also a question about being *fair* in relationship to the other stakeholders and thus showing a good *citizenship*.

The business developer of the municipality sees the value of the place image not only from the viewpoint of The Ironworks Village but more from that of the industrial activities of the place and the municipality and neighboring municipalities. He emphasizes the networks of the industrial organizations as a powerful engine for keeping the region alive and showing thus a *responsibility* for the whole region.

"A very significant customer group are the industrial customers, owners of the firms, partners, and workers. The local industry companies are the biggest customers for local accommodation and other service firms. It is the business of the traditional industrial companies that keeps the wheels turning at The Ironworks Village. [...] The industrial visitors, of course, they come on business, and are not there because of the environment or the Aalto buildings, but nevertheless The Ironworks Village is a place with a unique atmosphere based on more than 300 years of industrial history surrounded by a natural environment that makes it a pleasant place to stay on a business trip." (The manager of the business development agency)

The manager of the business development agency sees the potential in the segment of the industrial visitors and calls for more services for this specific group which according to him should not be ignored in the future development of the region. The fact that a reasonably small municipality like this has got so much industry is one of the most crucial factors for the wellbeing of the region.

"There are not so many tourists that would come to visit The Ironworks Village from farther away who would spend a night here. If we counted only on them, the situation would be much more desperate for the local service companies. Now the demand for the accommodation capacity is at a rather high-level thanks to the industrial businesses and its networks. But this means that you also must tailor your services to this segment of customers. This is the segment with the most potential for us. It is a big group, and it is a significant group for a place with an industrial character like this. Without the industry, the situation would be something quite different". (The manager of the business development agency)

The reasoning of the manager of the business development agency echoes with Williams and Aitken (2011, 449) emphasizing the co-creation of value to be about mutual dependency and reciprocal exchange.

4.3 Ethical Value Propositions Stemming from the Identity of a Place

Given the growing emphasis on interactive place branding across conventional, professional, and territorial boundaries, it is relevant to enter into a dialogic discourse. This view is in line with Kavaratzis and Hatch (2013) who not only base the conceptualization of place branding on the importance of stakeholders but view place brands as stemming from the place's identity, which is understood as an integrative process of identity construction through a dialogue between the stakeholders. The values (Rocha & Miles 2009) of the place were comprehensively identified by the personnel manager in the following citation in which she separates cultural and natural values and the values of respecting human beings. The identity of The Ironworks Village is expressed in the architecture of the buildings, and in the natural values of the lake and its surroundings which is labeled as a Natura region[1]. The humanistic values are expressed in the joint development projects with the idea of togetherness.

"The values of the place relate closely to cultural and natural values and respecting other human beings. For example, the buildings and the architects have been listed. Nature (the lake) comes close and it is protected by part of the protected Natura area with the flora, for example, rare tree species. Between human beings, it is essential if there would be a shared direction between the actors. The idea behind the developing project is also to achieve more togetherness". (Personnel manager)

The personnel manager is aware of the global and national trends of the growth in tourism. She connects this fact with the contemporary values of the people emphasizing wellbeing. She also sees the concept of wellbeing as changing according to people's experiences. This coincides well with the reasoning of Kavaratzis (2012) stating that the stakeholders of the brand-building process are constantly re-defining the culture, identity, and image of the place.

For the local micro-entrepreneurs, the identity of a place is reflected through their entrepreneurial identity stemming from the personal roots to the place and preferably from the family history. The owner of the perfume shop connects the values of the family history to the national and even to European history. The owner of the perfume shop has also inherited her mother's way of doing international business. One of the most important values that she has adopted in her business is that she appreciates the customer very much—for her the customer is the respectful queen to whom she offers her services. She considers the core idea of branding to be to differentiate oneself from the competitors.

"It is the service that differentiates a person from others and then the professionalism. I have studied the history of perfumes for 20 years. If you come and buy from me, you will learn a lot... about the past and Europe and the whole world." (Owner of the perfume shop)

The values of the place for the owner of the Design Sauna come from her childhood. Her father had worked at the factory and it was a dream come true for him when she and her brother bought the building and decided they would renovate it. After the renovation, they started a design café. The building designed by Alvar Aalto was built in the 1940s and it worked as a sauna (bathhouse) and laundry for the factory workers. Her uncle was employed by the factory and he was responsible for the sauna and the building. His family also lived in the house. The furniture in the Design Sauna is designed by Alvar and his wife and fellow architect, Aino Aalto. The owner of the café feels that the Aalto furniture collected by her brother has finally found a home. This reflects her loyalty to the cultural heritage of the place.

The entrepreneurs see the identity of the place as mirroring their entrepreneurial identity, whereas the representative of the municipality considers the identity from a broader perspective. The goal of the municipality is to enhance the wellbeing of the residents as well as develop and support business activities in a sustainable and economical manner. Therefore, it is important that both the municipality and other public stakeholders, entrepreneurs, and NGOs are aware of their own role and speak and act accordingly. However, the leading principle of the cooperation should be respectful humanity.

"We have to bear in mind each actor's role and business idea or task, for example, the municipality's basic role is to advance the wellbeing of its residents and the vitality of its areas and offer services for the residents. As to business companies, the economic aspect is emphasized. The actors have different business ideas, but we must also remember to appreciate each other. I think we've been able to apply an appreciative principle during this co-development process". (Personnel manager)

In sum, the interview excerpts of the stakeholders involved in the brand-building process of a rural community show how deeply rooted the personal values are in their doings. Naturally, the position of the person also signifies the importance of for example the place culture, place image, and place identity. The entrepreneurs recognized the importance of their family history, and the tangible relics of it in the form of buildings, furniture, or the documentary artifacts of business correspondence. The place culture was expressed in the way they identified with the place. The people who represented the municipality or other organizations saw the identity and the image of the place in a broader sense. They were more concerned about initiating coordination and communication between the actors as well as promoting wellbeing based on the place's cultural heritage (municipality), seeking partnerships, and networking opportunities (municipality). The excerpts represent positive brand attributes of the place, for example, the beautiful landscape (municipality), a long tradition (industry), roots, families (several actors), support of industrial activities (several actors), living in the area (inhabitants). Moreover, the industrial and architectural heritage of the place comes to light through the excerpts from the actors' narratives, as does the notion of cooperation and believing in a dream (whole network).

The theoretical place branding discussions that flow from Lemmetyinen et al. (2018) prompted us to focus on the ethical value propositions in place branding. Figure 5.2 below illustrates the ethical and human considerations of the place branding stakeholders co-operating in the brand-building process of a rural community.

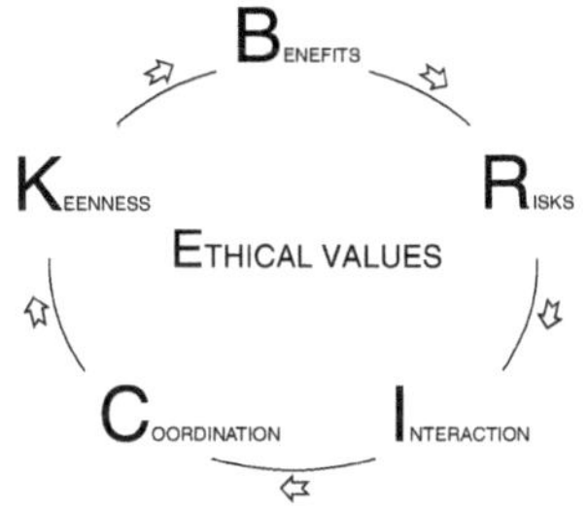

Figure 5.2 The ethical values complete the BRICKE model. Source: Author's own elaboration on the original in Lemmetyinen et al. (2018).

In prior studies (Lemmetyinen et al., 2018), benefits, risks, interaction, coordination and keenness have been found to represent the dimensions of stakeholder cooperation. Our comparisons of those classifications and the six universal moral values reveal similarities. Trustworthiness and caring would be associated with *benefit*, caring with *risks*, and respect with *interaction*, while citizenship would be included under *coordination* and fairness under *keenness*. The *ethics* component would also incorporate responsibility.

We identified values as being global (e.g., the environment or historical sites), regional (e.g., architecture) or personal (e.g., family relationships) in the research material. The prevalent view in this study makes it possible to incorporate a holistic ethical view of place branding. The major threat to establishing the ethical view is simply that it is omitted. The management has to make a conscious decision and provide direction to adopt an ethical approach as standard.

5 Conclusions

This research focuses on how a humanistic view of marketing is reflected in the place branding of a rural community offering value propositions in a stakeholder friendly and ethical manner. The ongoing dialogue between the stakeholders of the brand-building process as the value co-creators echoes the contemporary marketing discourse—service-dominant logic, which implicitly covers the humanistic view of management with its stakeholder focus. We followed Williams and Aitken (2011) and Stanislawski (2011) to make ethical values explicit (Schwartz 2005; Rocha & Miles 2009).

Our study followed the reasoning of Kavaratzis and Hatch (2013) setting the interaction between the place's culture, the place's identities, and the place's image at the heart of the conceptualization of place brand-building. In the identity-based approach of Kavaratzis and Hatch (2013) the culture, the image, and the identity of the place were interacting with the *ethical value propositions* of the stakeholders by adopting the *humanistic view on management and marketing.*

Based on the empirical evidence of the research the *humanistic view of the place marketing* can be identified in the activities of the individual stakeholders. We found that the value propositions of the stakeholders clearly stem from the culture of the place interwoven in the image and identity of the place as indicated in the model of identity-based place marketing. It was possible to identify values on a global, regional, or personal level in the interviews. Conscious humanistic management offers the potential to illuminate those values. This research contributes to the theory of place branding by adding the ethical value propositions and a humanistic view of the stakeholders as a novel perspective in the value co-creation of identity-based place branding. In addition, the context of the study, a rural community, offers a fresh view to the place branding theoretical discussion to produce ethical proposals on place branding.

Frequently mentioned limitations of a qualitative study include the issues of credibility, transferability, and dependability. In our study, *credibility* (cf. Marshall and Rossman, 1989; Gobo, 2004) accrues from following the case over several years and carefully documenting the accounts of the informants. Two of the authors represent the disciplines of marketing and entrepreneurship; and the third author, who was a personnel manager of the municipality involved, brought unique insights to the research. *Transferability* was strengthened through the triangulation of multiple respondents (representing various actor groups). We believe that if the study were to be replicated in the same context the results would be the same, thus confirming the *dependability* of the study. An interesting avenue for further studies would be the option of testing the BRICKE model in another context.

This study allows for the positioning of brands as a component of a social system designed to overcome provocations and present opportunities that leverage the potential of people as citizens, visitors, tourists, consumers, workers, artists, and co-producers of brands. Through the Kavaratzis and Hatch model, the ethical value propositions can be thematically classified with the help of the value model of M. Schwartz. This also supports the contention that the value theoretical result reported by S. Schwartz, in other words, that values are relatively unchanged between countries, and the differences of values are more likely between the individuals.

Note

1 https://ec.europa.eu/environment/nature/natura2000/biogeog_regions/index_en.htm.

References

Aaker, D. (2014). *Aaker on branding: 20 principles that drive success*. New York: Morgan James Pub.

Aaker, D. (2004). *Brand portfolio strategy: creating relevance, differentiation, energy, leverage, and clarity*. New York: The Free Press.

Aaker, D. & Joachimsthaler, E. (2000). *Brand leadership*. New York: The Free Press.

Aaker, D. (1991). *Managing Brand Equity: Capitalizing on the Value of a Brand*. New York: The Free Press.

Beverland, M., Lindgren, A., Napoli, J., Ballantyne, D. & Aitken, R. (2007). Branding in B2B markets: insights from the service-dominant logic of marketing. *Journal of Business & Industrial Marketing*, *22*(6), 363–371.

Cieciuch, J. & Schwartz, S. H. (2012). The Number of Distinct Basic Values and Their Structure Assessed by PVQ-40. *Journal of Personality Assessment 2012*, *94*(3), 321–328.

Coffey, A. & Atkinson, P. (1996). *Making Senseof Qualitative Data*. London: Sage.

Datzira-Masip, J. & Poluzzi, A. (2014). Brand architecture management. The case off our tourist destinations in Catalonia. *Journal of Destination Marketing & Management*, *3*(1), 48–58.

Fisk, R. P. (2019). Commentary on "Humanistic Management of Social Innovation in Service (SIS): An Interdisciplinary Framework". *Humanistic Management Journal*, *4*(2), 187–189.

Freeman, R. E. (1984). *Strategic Management: A Stakeholder Approach*. Boston, MA: Bitman.

Gobo, G. (2004). Sampling, representativeness and generalizability. In Seale, C., Gobo, G., Gubrium, J. F. & Silverman, D. (Eds.), *Qualitative research practice* (pp. 435–456). London: Sage.

Hatch, M. J. & Schultz, M. (2002). The Dynamics of Organizational Identity. *Human Relations*, *55*(8), 989–1018.

Kabadayi, S., Alkire, L., Broad, G. M., Livne-Tarandach, R., Wasieleski, D. &

Puente, A. M. (2019). Humanistic Management of Social Innovation in Service (SIS): an Interdisciplinary Framework. *Humanistic Management Journal*, *4*, 159–185.

Kant, I. (1997). *Critique of Practical Reason*. New York: Cambridge University Press.

Kavaratzis, M. & Ashworth, G. (2008). Place Marketing: How did we get here and where are we going? *Journal of Place Management and Development*, *1*(2), 150–167.

Kavaratzis, M. (2012). From "necessary evil" to necessity: stakeholders' involvement in place branding. *Journal of Place Management and Development*, *5*(1), 7–19.

Kavaratzis, M. & Hatch, M. J. (2013). The dynamics of place brands: An identity-based approach to place branding theory. *Marketing Theory*, *13*(1), 69–86.

Kramer, M. & Pfitzer, M. (2016). The Ecosystem of Shared Value. *Harvard Business Review*, *94*(10), 80–89.

Lemmetyinen, Arja, Nieminen, Lenita & Aalto, Johanna, (2018). Who takes the lead in initiatingcooperation in a cultural network and why? A case of rural Finnish destination. In Egberts, L. & Alvarez Maria, D. (Eds.), *Heritage and Tourism: Places, Imageries and the Digital Age*. Amsterdam University Press, 95–114.

Lemmetyinen, A. & Go, F. M. (2010). Building a brand identity in a network of Cruise Baltic's destinations. A multi-authoring approach. *Journal of Brand Management*, *17*(7), 504–518.

Lindgreen, A., Palmer, R., Wetzels, M. & Antioco, M. (2009). Do different marketing practices require different leadership styles? An exploratory study. *Journal of Business & Industrial Marketing*, *24*(1), 14–26.

Lovins, L. H. (2016). Needed: A Better Story. *Humanist Management Journal*, *1*(1), 75–90.

Luhmann, N. (1986). The Autopoiesis of Social Systems. In Geyer, F. & Zeuwen, J. Van d. (Eds.), *Sociocybernetic Paradoxes: Observation, Control and Evolution of Self-Steering Systems* (pp. 172–192). London: Sage.

Luhmann, N. (2012). *Theory of Society, volume 1*. Translated by Barret, Rhodes. Stanford: Stanford University Press.

Mabey, C. & Freeman, T. (2012). Four Readings of Place and Brand Leadership. In Go, F. M. & Govers, R. (Eds.), *International Place Branding Yearbook Managing Smart Growth and Sustainability* (pp. 33–44). Basingstoke: Palgrave MacMillan.

Marshall, C. & Rossman, G. B. (1989). *Designing qualitative research*. London: Sage Publications.

Moilanen, T. & Rainisto, S. (2009). *How to brand nations, cities and destinations: a planning book for place branding*. Basingstoke: Palgrave Macmillan.

Porter, E. & Kramer, M. (2011). The Big Idea: Creating Shared Value. How to Reinvent Capitalism – and Unleash a Wave of Innovation and Growth. *Harvard Business Review*, *89*(1-2), 62–77.

Riessman, C. (2004). Narrative interviewing, In Lewis-Beck, M. S., Bryman, A. & Futing, L.T. (Eds.), *Encyclopedia of Social Science Research Methods* (pp. 705–709). London: Sage Publications.

Rocha, H. & Miles, R. (2009). A Model of Collaborative Entrepreneurship for a More Humanistic Management. *Journal of Business Ethics*, *88*, 445–462.

Schwartz, M. (2005). Universal Moral Values for Corporate Codes of Ethics. *Journal of Business Ethics*, *59*(1-2), 27–44.

Schwartz, S. H. (1992). Universals in the Content and Structure of Values: Theory and Empirical Tests in 20 Countries. In Zanna, M. (Ed.), *Advances in Experimental Social Psychology*, 25, (pp. 1–65). New York: Academic Press.

Schwartz, S. H. (2014). Societal Value Culture: Latent and Dynamic. *Journal of Cross-Cultural Psychology*, *45*(1), 42–46.

Spitzeck, H. (2011). An Integrated Model of Humanistic Management. *Journal of Business Ethics*, *99*, 51–62.

Stanislawski, Sumire, (2011). The Service-dominant Logic of Marketing and the Ethics of Co-Creation. *The Bulletin of the Graduate School of Commerce – Waseda University*, *73*, 109–133.

Tedlock, B. (2000). Ethnography and ethnographic representation. In Denzin, N.K. & Lincoln, Y.S. (Eds.), *Handbook of Qualitative Research* (pp. 455–488). Thousand Oaks: Sage Publications.

Vargo, S. L. & Lusch, R. F. (2004). Evolving to a New Dominant Logic for Marketing. *Journal of Marketing*, *68*(1), 324–335.

Vargo, S. L. & Lusch, R. F. (2008). Service-dominant Logic: Continuing the Evolution. *Journal of the Academy of Marketing Science*, *36*(1), 1–10.

Warnaby, G. (2009). Towards a service-dominant place marketing logic. *Marketing Theory*, *9*(4), 403–423.

Williams, J. & Aitken, R. (2011). The Service-Dominant Logic of Marketing and Marketing Ethics. *Journal of Business Ethics*, *102*(3), 439–454.

6 Service Delivery and Employee Turnover

The Importance of Work-Life Balance

Giuseppe Cappiello, Gabriele Morandin, and Manuela Presutti

1 Introduction

According to the most influential perspectives, such as Resource-Based Theory (Barney, 1991) and Strategic Human Resource Management (Su, Wright, & Ulrich, 2018; Wright & McMahan, 1992), employees represent one of the firm's main strategic assets, especially in services (Heskett, Sasser, & Schlesinger, 1997; Kusluvan, Kusluvan, Ilhan, & Buyruk, 2010; Nieves & Quintana, 2018).

This claim contrasts with the evidence that the service industry is characterized by extremely high turnover. Indeed, it is common for workers to change jobs often. In the "accommodation and food services" sector, for instance, the turnover rate was 58.3% in 2019, with an increase of about 10% in 5 years (US Bureau of Labour Statistics, 2020).

For that reason, staff turnover represents a big challenge in service management, and tourism in particular (Li, Ma, & Qu, 2017), due to its direct impact on costs (i.e., recruiting, selecting, and hiring, or costs from error, supervisory disruptions, and vacancies) and profitability (Hinkin & Tracey, 2000; Tracey & Hinkin, 2008). In a service organization, where human encounters are central, it seems reasonable to expect that employee dissatisfaction and high levels of turnover will be associated with lower levels of customer satisfaction (Hurley & Estelami, 2007). Companies that satisfy employees, by contrast, can preserve a better work environment, promote greater customer satisfaction, and enhance financial performance and organizational effectiveness (Chi & Gursoy, 2009; Koys, 2001; Kumar, Jauhari, Rastogi, & Sivakumar, 2018). When people depart, their know-how regarding processes, methods, and customers also leaves, dispersing a rich source of knowledge that takes time to cultivate.

Although much has already been written about turnover (Hom, Lee, Shaw, & Hausknecht, 2017), empirical research on this theme from a humanistic perspective is, to the best of our knowledge, still scarce. Humanistic Management "is oriented not only to obtaining results through people, but also, and above all, toward people themselves, showing care for their flourishing and well-being" (Melè, 2016, p. 33). Previous studies have long tried to identify the determinants of the intention to change jobs, as well as the various initiatives that may promote employment retention (Griffeth, Hom, & Gaertner, 2000), but rarely has the problem been tackled from another point of view. Studies on tourism and hospitality often prioritize the company's perspective and profitability over workers' personal needs. Nonetheless, changing jobs has important personal repercussions. Although this event might entail psychological (disappointing expectations) or economic (foregoing opportunities) sacrifices, some people do leave their jobs in order to improve their quality of life or better align their work with their values and interests (e.g., starting a family; Deery & Jago, 2009).

In order to broaden our understanding of turnover and seasonal employment in service delivery, we ran an exploratory analysis in a well-known seaside resort in central Italy (Rimini). We chose this location because tourists' presence is limited to the summer period and other moments in conjunction with fairs or other events; as such, the workforce does not have the chance to work for the same company in the winter, as there are few opportunities in this area. Moreover, tourism in Italy is facing a human resource shortage that is hindering businesses' ability to address peak demand.

In short, this chapter investigates the turnover occurrence according to a Humanistic Management approach (Pirson & Kimakowitz, 2014; Pirson & Lawrence, 2010).

In this chapter, we first present a review of the relevant literature and then follow with a description of our research, the results, and the main implications.

The results of this exploratory study indicate that commitment and job satisfaction are determinants of the intention to remain with the same company, whereas organizational identification has a significant effect on commitment. Paradoxically, the economic requirement is considered less important; one of the main causes behind a decision to change jobs is work-life balance, that is, the possibility of reconciling work obligations with the rest of one's life.

An improved work-life balance not only leads to better short-term outcomes for the company but also helps organizations and employees become more sustainable over time and harmonious with the surrounding community.

2 Theoretical Framework

One of the factors that put a strain on service management and tourism businesses, in particular, is seasonality: the significant variation in demand across periods. This period may be a month (as it is for hospitality and exhibitions) or a day (as it is for restaurants, car rentals, or food delivery). This issue is becoming more prevalent thanks to the globalization and the expansion of transportation options, which facilitate different movement patterns, and the economic crisis, which has helped reduce the amount of time that people spend away from home. In order to properly respond to this situation, tourism organizations need to establish a structure that can withstand peak demand, as well as manage reservations in a way that spreads requests and maximizes revenues for each period and service category.

Moreover, tourism companies are labor-intensive: Some roles are fixed (i.e., accountants, security, maintenance, and many others) and cannot be modified to suit fluctuating customer quantities, whereas a portion of the staff is hired by forecasting a rise in demand.

This means that the working hours can be very different and stressful according to the periods, with shifts sometimes longer than normal. Companies hire a considerable share of workers intermittently to meet temporary needs, which entails limited career prospects. Such workers include low-qualified people, students, internships, migrant workers, and persons looking to supplement their existing income, or those who rely on temporary employment while looking for more stable positions (weekends or peak months). The workforce is an especially critical aspect of this sector and, in many countries, it is difficult to retain talents or sometimes even find the necessary staff to run the business (Deery, 2008; Stamolampros, Korfiatis, Chalvatzis, & Buhalis, 2019).

In this situation, it is common for workers to move from one firm to another in search of a more stable position, which can have significant repercussions in terms of disappointed expectations, personal sacrifices, and quality of life. Iverson and Deery (1997) call “turnover culture” the “acceptance of the turnover as a part of the work-group norm” (p. 71). Their research uncovered that, in many cases, workers already know that their employment will last a short time because that is customary in the sector.

Given that the tourism field now tacitly accepts a low involvement and a high rate of turnover, it is important to consider how this issue might lower quality standards and customer satisfaction while raising costs and eroding knowledge for the company and territory.

Of course, employee turnover is a long-standing concern. As Hom et al. (2017) review of 100 years of studies indicates, scholars have spent considerable time trying to understand the phenomenon, the methods to analyze them, and the possible trajectories of future research. In 1917, as an example, Fisher was already looking for “how to reduce labor turnover” as society was becoming more industrialized, spurring profound changes in work habits and lifestyles. At that time, the main problem seemed to be the costs of turnover, which used to be estimated as double a worker’s annual salary.

It is impressive to think that nowadays, artificial intelligence is the primary method of identifying individuals who are most likely to remain in a position. Generally speaking, turnover occurs when there is an interruption of working ties, which can happen for numerous reasons.

Maertz and Campion (2004) divide research on turnover into two categories: process models and content models. "Process models focus on *how* individuals arrive at their final decisions to quit, while content models focus on *why* individuals quit organizations" (Maertz & Campion, 2004, p. 566). The reasons why a worker decides to stop work can be various. Personal characteristics (gender, race, and cognitive abilities) seem to have no predictive force, whereas job satisfaction, organizational commitment, and comparable alternatives strongly influence the decision. Other dimensions of work experience (participative management, promotional chances, work group cohesion, role stress, pay, pay satisfaction, distributive justice, and job scope) demonstrate a smaller effect (Griffeth, Hom, & Gaertner, 2000).

Through a path analysis, Hurley and Estelami (2007) found that employee satisfaction is driven by many factors such as training, scheduling, communication, empowerment, conflict resolution, etc., but the relation is non-linear.

Other studies from the restaurant sub-sector highlight that the perceived warmth and competence of co-workers is a determinant for job satisfaction (Bufquin, Di Pietro, Orlowski, and Partlow 2017), whereas customer incivility has a positive effect on turnover intention (Han, Bonn, & Cho, 2016).

Wasmuth and Davis (1983) demonstrated that supervision conditions and pay lead to dissatisfaction. Van Dick et al. (2004) proved that a strong organizational identification and a low turnover intention are correlates of dissatisfaction, whereas employees' satisfaction acts as a mediator.

Many scholars (i.e., Stanley, Vandenberghe, Vandenberg, & Bentein 2013; Weng & McRoy, 2012) assert that commitment plays a primary role. Commitment is defined as the psychological state of the professional relationship and has significant implications for desirable attitudes and behavior.

Cho, Johanson, and Guchait (2009) add that perceived supervisor support, perceived organizational support, and organizational commitment affect the intention to leave. These authors also verified that the intention to leave and the intention to stay have not equal antecedents.

Zimmerman, Swider, and Boswell (2019), seeking to "[establish] a parsimonious content model that succinctly captures the important, unique aspect of multiple existing models" (p. 100), settled on five factors that can predict employee turnover: effect toward the organization, work environment, instrumental attachment, extra organizational ties, and sense of obligation.

Delery and Doty (1996) found that an effective Human Resources (HR) office and policy are key factors in reducing turnover. HR should focus on employee retention and development by providing a supportive work environment and improving the perception that employees have opportunities to grow inside the firm. Consequently, more competent and loyal employees benefit the organization through higher service quality. In the same vein, HR should implement a strong communication system (Labov, 1997) that allows employees to share information, participate in decision-making, and foster a positive corporate culture (Meaghan & Bontis, 2002). These efforts can lead to a comfortable environment in which personnel better identify with their job and work more efficiently.

Mitchell, Holtom, Lee, Sablynski, and Erez (2001) introduced a new construct, job embeddedness, and therefore a new perspective, to reduce turnover:

> Being embedded in an organization and a community is associated with reduced intent to leave and reduced actual leaving. These findings appear to support the current emphasis in the academic and popular press on the need for organizations to be concerned with employees' lives both on and off the job. It also suggests that a focus on money and job satisfaction as the levers for retention may be too limited. (p. 1118)

In adopting the Humanistic Management approach, which "[aims] for a more complete image of the person" (Lutz & Lux, 1988), we seek to deepen this intuition and look for possible answers in the hospitality sector.

In brief, the Humanistic Management movement arose in the 21st century as an alternative to a mechanistic vision of economic activity, instead emphasizing the role of people and their goals rather than the techniques adopted to achieve better performance (Melè, 2016).

As previously mentioned, tourism businesses commonly impose very long shifts over 7 days and holiday seasons. Furthermore, workers face significant pressure from their continuous contact with the customer. Cullen and McLaughlin (2006) argue that hotel managers risk suffering from "presenteeism" because they think they are obligated to represent the hotel all day long and provide constant support to customers and staff. These industry-specific characteristics make it difficult to balance private life with work obligations.

Sirgy and Lee (2018, p. 229) affirm that work-life balance "plays an important role in individual well-being such as health satisfaction, family satisfaction and overall life satisfaction". It specifically involves two dimensions: role engagement in work and nonwork life, and minimal conflict between work and nonwork roles.

Although the antecedents of work-life balance may change based on the context (i.e., the culture of reference), some aspects are cited frequently across various countries, such as long working hours, time pressure, lack of flexibility, job insecurity, role clarity, job autonomy, or supervisory support (Haar, Sune, Russo, & Olierre-Malaterre, 2019). Moreover, the objective problems and psychological tensions of reconciling private life and work-related requests are reflected not only in efficiency at work but also in family relationships and time spent out of work (Deery & Jago, 2009).

In conclusion, we seek to interpret the phenomenon of staff turnover in line with a Humanistic Management view, focusing especially on the aspects of workers' private lives.

3 Methodology

3.1 Procedures, Setting, and Sample

In order to understand some fundamental aspects of seasonal employment, the determinants of turnover, and the local perception of the phenomenon, we started our research with two focus groups: one interacting with tourism entrepreneurs and the other with representatives from trade unions and local institutions. The first group included both seasonal and non-seasonal entrepreneurs so that it would be less affected by the winter interruption. We began each focus group by asking participants to describe "critical incidents" (Flanagan, 1954) or actual events related to seasonal work that had caused problems for management, employment relationships, and/or service delivery. Based on the literature review and focus groups, we identified four domains for our investigation: work motivation, organizational identification, commitment, and job satisfaction.

We then administered a questionnaire at the provincial offices in Rimini, a famous seaside resort in Central-North Italy. We chose this method of collecting the questionnaires because all workers must deliver documents to the office before starting work. A total of 619 seasonal workers completed the questionnaire. The sample was 54.4% female, in line with population, with an average age of 24.09 years old (standard deviation equals to 11.2, a little younger than population).

3.2 Measures

To identify the antecedents of the intention to leave the job, we used a 31-item scale developed by Morandin and Bergami (2014) to assess temporary workers' work motivation (Table 6.1).

Respondents expressed their agreement to each sentence on a 7-point Likert scale. By applying an exploratory factor analysis to this scale, we uncovered four factors that were labeled: Desire to grow (e.g., skill development, personal growth), relational needs (e.g., socialization, relations

Table 6.1 Work motivation

I Chose This Job	Mean	St. Dev.
Desire to grow	**4.6**	**1.2**
1. To enhance my career prospects	4.1	1.9
2. For self-fulfillment	4.6	1.8
3. For personal growth	5.4	1.7
4. Because will allow me to continue my education	4.1	1.9
5. To find new motivation	4.4	1.8
6. For personal satisfaction	5.4	1.7
7. To have a better life	4.3	1.8
8. To satisfy the company's customers	4.0	1.9
9. To test myself	5.2	1.7
10. For my personal characteristics (so I am)	4.2	1.9
11. For more psychological autonomy	4.4	1.8
12. To enrich my curriculum vitae	5.0	1.8
Relational needs	**4.7**	**1.1**
13. To establish relations with customers	4.5	1.8
14. To socialize with other people	5.0	1.6
15. To meet new people	4.6	1.7
16. To feel useful to others	4.9	1.7
17. Because I wanted a new experience	4.9	1.8
18. Because it was the first opportunity that I have found	3.9	2.0
Economic needs	**5.8**	**1.0**
19. To set aside a bit of money	6.0	1.4
20. To be more independent from an economic point of view from my family	6.2	1.2
21. Because it gives me the opportunity to earn	5.5	1.6
22. Because it allows me to pay my way	5.5	1.5
23. In order to satisfy my needs and desires	5.6	1.5
Goals of work-life balance	**4.2**	**1.1**
24. Because it allows a balance between working life and private life	4.1	2.0
25. Because some friends suggested me to accept the proposal of working	3.1	2.0
26. Because there is a good working environment	4.9	1.8
27. Because the workplace is close to home	4.2	2.2
28. Because the company has a good reputation	4.5	1.8
29. For increased stability	5.0	1.6
30. Because it is a part-time job	3.1	2.1
31. Because the work is well suited to my profile	4.6	2.4
Min: 1–Max: 7		

Source: Author's own elaboration based on Morandin and Bergami (2014)

with clients), economic requirements (e.g., paying expenses, saving money), and work-life balance (e.g., part-time job, workplace close to home).

We then measured organizational identification through a visual scale developed by Bergami and Bagozzi (2000). Next, we used the 22-item scale developed by Meyer and Allen (1997) to measure the three main types of commitment: affective ("I would be very happy to spend the rest of my career with this organization" – indicating a feeling of attachment and positive emotion), continuance ("I feel that I have too few options to consider leaving this organization" – indicating a lack of alternatives), and normative ("I would feel guilty if I left my organization now" – indicating a moral obligation). Participants were invited to express their agreement on a 5-point Likert scale.

Job satisfaction was measured with a three-item measure. Another 11 items were added to investigate specific aspects of the job (e.g., supervision, conditions of work). Participants were invited to express their agreement on a 7-point Likert scale (Table 6.2).

Finally, we measured the intention to remain in the organization with a four-item scale adapted from Hom and Griffeth (1991). An example item is "If I could, I would stay in this organization for a long time". Respondents rated this item on a 7-point Likert scale.

4 Results and Discussion

4.1 Descriptive Statistics

Table 6.3 provides descriptive statistics and correlation among constructs. Although the location we sampled from has a long and prestigious

Table 6.2 Job satisfaction

Item	Media	Dev. Std.
I am satisfied with the relationship with my colleagues	5.3	1.6
In general, I'm satisfied with my work	5.2	1.6
I am satisfied with the relationship with my manager	5.1	1.7
I am satisfied with the quality of service that we realize	5.1	1.7
I am satisfied with the safety of my job	5.0	1.6
I am satisfied with my general working conditions	5.0	1.5
I am satisfied with the quality of the contribution of my manager	4.9	1.7
I am satisfied with the quality of the contribution of my colleagues	4.7	1.8
I am satisfied that I receive financial remuneration	4.6	1.8
I am satisfied with the other non-economic incentives that I receive (rewards, motivation, and opportunity)	4.6	1.9
I am satisfied with my opportunities for advancement	4.2	1.8
I am satisfied with the challenges that I find on the farm	4.2	1.8
I am satisfied with the management of this company	4.2	1.7
I am satisfied with my free time	3.8	2.1
Min: 1–Max: 7		

Source: Author's own elaboration based on Meyer and Allen (1997).

Table 6.3 Descriptive statistics and correlation among constructs

	Mean	Dev. St	1.	2.	3.	4.	5.	6.	7.	8.	9.	10.
1. Desire to grow	4.58	1.20	1									
2. Relational needs	4.66	1.14	0.616**	1								
3. Economic needs	5.76	1.01	0.378**	0.360**	1							
4. Work-life balance	4.18	1.11	0.508**	0.533**	0.291**	1						
5. Organizational identification	4.69	1.91	0.387**	0.284**	0.162**	0.356**	1					
6. Affective commitment	2.67	1.07	0.460**	0.341**	0.129**	0.447**	0.531**	1				
7. Continuance commitment	2.72	1.02	0.037	0.108*	0.064	0.118*	–0.038	0.151**	1			
8. Normative commitment	3.28	1.11	0.401**	0.295**	0.220**	0.371**	0.447**	0.594**	0.198**	1		
9. Job satisfaction	4.71	1.24	0.549**	0.411**	0.294**	0.530**	0.488**	0.623**	0.024	0.577**	1	
10. Intention to remain	4.21	1.78	0.498**	0.344**	0.213**	0.474**	0.422**	0.666**	0.124**	0.490**	0.673**	1

Source: Author's own elaboration.

** Correlation is significant at the 0.01 level (2-tailed); * correlation is significant at the 0.05 level (2-tailed).

tradition in the tourism and hospitality industry, our analyses made clear that respondents did not feel much professional motivation toward their work. Specifically, they seemed to envision their working activities as temporary and almost exclusively aimed at fulfilling monetary goals.

Regarding identification with the company, the respondents registered a mean value of 4.7, or an almost "moderate overlap". The standard deviation was 1.9, which indicates a degree of variability in the responses.

The results revealed that the predominant type of commitment is normative (average of 3.3). The average value of continuance commitment lies below the midpoint of the scale (average of 2.7), which is significant, as this value is better when low. However, the same is not true of the affective dimension andits fairly low average of 2.7 indicates clear room for improvement.

The average value of job satisfaction was 4.7: a value that is above the midpoint of the scale, but not particularly high (considering that the scale ranges from 1 to 7). Interestingly, respondents felt more satisfied with their relationships with colleagues and managers than with the work and quality of service they provide. Respondents expressed the greatest dissatisfaction with their leisure and career opportunities, which is an interesting insight for companies seeking to increase performance by improving employees' all-round experience.

Finally, the intention to remain was slightly above the average of the scale (mean equal to 4.2), showing a moderate intention to remain in the organization.

4.2 Structural Model and Its Discussion

Due to the multiple causal relationships being investigated in this research, we decided that a structural equation model was the most appropriate analytical approach (Jöreskog & Sörbrom, 2003). This model proposes organizational identification as a mediator of the effects that work motivation exerts on the three different facets of workers' commitment toward the employer (Figure 6.1). It should be noted that the structural equation model implemented with LISREL 8.80 perfectly fits the data (NFI, NNFI = 0.95, CFI, IFI = 0.96, GFI = 0.82; AGFI = 0.79, SRMR = 0.07, RMSEA = 0.07).

The main driver of organizational identification appears to be the factor "work-life balance", as suggested by the high value of the associated coefficient ($\gamma = 0.77$, $p < 0.001$). When looking at the specific questions that underlie this latent construct, we found that the most concrete elements of work-life balance are the working environment, the company's reputation, and the consistency between the actual work and the individual's profile. In practice, increasing individuals' perceptions of these aspects increases their identification.

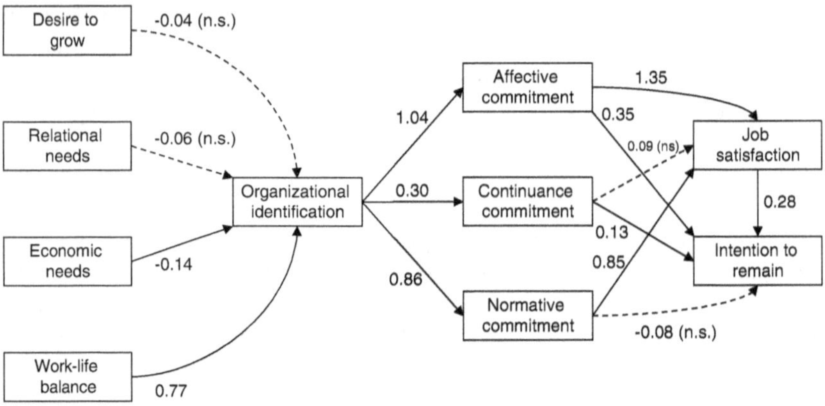

Figure 6.1 Predictors of the intention to remain.
Source: Author's own elaboration.

Economic needs also influence organizational identification, albeit to a lesser extent than work-life balance. In particular, the negative sign for the "economic needs" construct coefficient ($\gamma = -0.14$, $p = 0.02$) suggests that workers tend to identify less with an organization when the motivation to work has been primarily economic. Meanwhile, we did not find any statistically significant impact of desire for professional growth ($\gamma = 0.04$, $p = 0.65$) or relational needs ($\gamma = -0.06$, $p = 0.39$) on organizational identification. The R^2 value indicates that, in this model, 66% of the variance for the organizational identification variable ($R^2 = 0.66$) can be explained by its linear dependence on the four work motivation factors.

Moreover, the data show that organizational identification significantly affects all three types of commitment, albeit with different intensity: organizational identification seems to have the greatest impact on affective commitment ($\beta = 1.04$, $p < 0.001$), followed by normative commitment ($\beta = 0.86$, $p < 0.001$), and continuative commitment ($\beta = 0.30$, $p < 0.001$). In other words, the more that employees identify with the organization, the higher their level of affective and emotional attachment to the company (affective commitment), and the greater their feeling that such an attachment is the right thing to do (normative commitment). Notably, affective and normative commitments are not simply relevant constructs; they also directly and significantly impact job satisfaction, with beta values, respectively, equal to 1.35 ($p < 0.001$) and 0.85 ($p < 0.001$).

In line with previous research (Rust, Stewart, Miller, & Pielack, 1996), we found that job satisfaction positively affects employees' intention to remain with the company ($\beta = 0.28$, $p < 0.001$). Beyond corroborating the extant knowledge on the satisfaction–retention relationship, our results also show that intention to remain is directly influenced by affective commitment ($\beta = 0.35$, $p < 0.001$) and continuance commitment ($\beta = 0.13$, $p < 0.001$). In other words, we found that affective commitment exerts an effect on intention to remain that is partially mediated by job satisfaction. Meanwhile, job satisfaction fully mediates the normative commitment–intention to remain relationship, whereas continuance commitment fully mediates the identification–intention to remain relationship (granted, without affecting employees' job satisfaction).

In a nutshell, our results highlight the pivotal role of organizational identification in translating seasonal employees' motivation to work to their perceptions and future intentions. Specifically, jobs that offer a better work-life balance seem to enjoy a higher level of identification, which then generates higher levels of commitment. The affective and normative sides of this commitment can then positively shape seasonal employees' degree of satisfaction with their jobs.

5 Conclusions

In the hospitality industry, human resources are essential to delivering services and satisfying customers. However, Humanistic Management highlights that "little attention was paid to the human condition of workers, except for making them more efficient in productive operations" (Melè, 2003, p. 77). In that vein, the present research aimed to provide a deeper understanding of seasonal employment and the turnover phenomenon, especially because they are so critical to both the hospitality industry's success and the well-being of employees.

Our results indicate that employees' work motivation influences job satisfaction and the decision to remain with the company over time; however, the path from motivation to the intention to remain is not direct, but rather mediated by a set of relevant constructs. Work motivation affects the extent to which employees develop high levels of identification with the organization they work for. Moreover, in this particular industry, employees' organizational identification is largely driven by their work-life balance; as the latter increases, so does the former. Our results underscore that it is not the prospect of a future career nor the economic conditions that determine these workers' fondness toward their organizations.

Our results generally support Wasmuth and Davis' (1983) assertion that managerial practice, effective supervision, and sensitivity to individual aspirations, all inexpensive to adjust, can greatly improve employee morale. Consequently, touristic organizations should promote a positive working climate and leave more space for private life. This can be implemented, for example, by offering shifts that are more balanced or providing workers with facilities that help them in their private affairs. Job embeddedness has an effect on work and family conflict and the intention to quit the job (Treuren, 2019). Conversely, employees who feel like their relationship with the firm is purely economic may develop lower levels of organizational identification.

Current evidence (Stamolampros, Korfiatis, Chalvatzis, & Buhalis, 2019) suggests that businesses can lower turnover rates by providing adequate training and choosing the right recruiting strategies, but above all, they need to establish a communication channel with employees. Analyzing thousands of reviews left by workers on an online platform, the authors found that workers ask to be treated with greater respect from their managers.

Given that it would be unfair and insufficient to choose employees based on their economic necessities (or lack thereof), our findings suggest that managers should instead reduce their emphasis on monetary gains and focus on the relationships and quality of life offered by the work. In other words, employees should be informed about the job's advantages in terms of life quality from the beginning of the hiring process. On this point, policymakers should also carefully monitor seasonal employees' working hours and unbecoming behavior under the profile of work-life balance because they both have the potential to lower individuals' organizational identification, thereby decreasing their productivity and, more generally, the service quality of the tourism industry as a whole.

Acknowledgments

We want to thank Prof. Gabriele Pizzi for the useful suggestions and the Provincial Labor Office and Cescot for the availability of the data. Obviously, the responsibility belongs only to the authors. The authors contributed equally to the chapter.

References

Barney, J. (1991). Firm resources and sustained competitive advantage. *Journal of Management*, *17*(1), 99–120. doi:10.1177/014920639101700108.

Bergami, M., & Bagozzi, R. P. (2000). Self-categorization, affective commitment and group self-esteem as distinct aspects of social identity in the organization. *British Journal of Social Psychology*, *39*(4), 555–577. doi.org/10.1348/014466600164633.

Bufquin, D., Di Pietro, R., Orlowski, M., & Partlow, C. (2017). The influence of restaurant co-workers' perceived warmth and competence on employees' turnover intentions: The mediating role of job attitudes. *International Journal of Hospitality Management*, *60*, 13–22. doi.org/10.1016/j.ijhm.2016.09.008.

Chi, C., & Gursoy, D. (2009). Employee satisfaction, customer satisfaction, and financial performance: An empirical examination. *International Journal of Hospitality Management*, *28*(2), 245–253. doi.org/10.1016/j.ijhm.2008.08.003.

Cho, S., Johanson, M., & Guchait, P. (2009). Employees intent to leave: A comparison of determinants of intent to leave versus intent to stay.

International Journal of Hospitality Management, *60*, 374–381. doi.org/10.1016/j.ijhm.2008.10.007.

Cullen, J., & McLaughlin, A. (2006). What drives the persistence of presenteeism as a managerial value in hotels? Observations noted during an Irish work-life balance research project. *International Journal of Hospitality Management*, *25*(3), 510–516.

Deery, M. (2008). Talent management, work-life balance and retention strategies. *International Journal of Contemporary Hospitality Management*, *20*(7), 792–806. doi.org/10.1108/09596110810897619.

Deery, M., & Jago, L. (2009). A framework for work-life balance practices: Addressing the needs of the tourism industry. *Tourism and Hospitality Research*, *9*(2), 97–108. doi.org/10.1057/thr.2009.4.

Delery, J. E., & Doty, D. H. (1996). Modes of theorizing in strategic human resource management: Tests of universalistic, contingency, and configurational perspectives. *Academy of Management Journal*, *39*, 802–835.

Fisher, B. (1917). How to reduce labour turnover. *The Annals of the American Academy of Political and Social Science*, *17*, 10–32.

Flanagan, J. C. (1954). The critical incident technique. *Psychological Bulletin*, *51*(4), 327–358. doi.org/10.1037/h0061470.

Griffeth, R. W., Hom, P. W., & Gaertner, S. (2000). A meta-analysis of antecedents and correlates of employee turnover: Update, moderator tests, and research implications for the next millennium. *Journal of Management*, *26*(3), 463–488. doi.org/10.1177/014920630002600305.

Haar, J. M., Sune, A., Russo, M., & Olierre-Malaterre, A. (2019). A cross-national study on the antecedents of work–life balance from the fit and balance perspective. *Social Indicator Research*, *142*, 261–282. doi.org/10.1007/s11205-018-1875-6.

Han, S. J., Bonn, M. A., & Cho, M. (2016). The relationship between customer incivility, restaurant frontline service employee burnout and turnover intention. *International Journal of Hospitality Management*, *52*, 97–106. doi.org/10.1016/j.ijhm.2015.10.002.

Heskett, J., Sasser, W. E., Jr., & Schlesinger, L. (1997). *The service profit chain: How leading companies link profit and growth to loyalty, satisfaction and value*. New York: Free Press.

Hinkin, T., & Tracey, J. B. (2000). The cost of turnover. *Cornell Hotel and Restaurant Administration Quarterly*, *41*(3), 14–21. doi.org/10.1177/001088040004100313.

Hom, P. W., & Griffeth, R. W. (1991). Structural equations modeling a test of turnover theory: Cross-sectional and longitudinal analyses. *Journal of Applied Psychology*, *76*, 350–366.

Hom, P., Lee, T. W., Shaw, J. D., & Hausknecht, J. P. (2017). One hundred years of employee turnover theory and research. *Journal of Applied Psychology*, *102*(3), 530–545. doi:10.1037/apl0000103.

Hurley, R., & Estelami, H. (2007). An exploratory study of employee turnover indicators as predictors of customer satisfaction. *Journal of Services Marketing*, *21*(3), 186–199. doi: 10.1108/08876040710746543.

Iverson, R., & Deery, M. (1997). Turnover culture in the hospitality industry.

Human Resources Management, 7(4), 72–81. doi.org/10.1111/j.1748-8583.1997.tb00290.x.

Jöreskog, K. G., & Sörbrom, D. (2003). *Lisrel 8.7 for Mac Os 9 and X*. Lincolnwood, IL: Scientific Software International.

Koys, D. J. (2001). The effects of employee satisfaction, organizational citizenship behavior, and turnover on organizational effectiveness: A unit-level, longitudinal study. *Personnel Psychology*, *54*, 101–114. doi.org/10.1111/j.1744-6570.2001.tb00087.x.

Kumar, M., Jauhari, H., Rastogi, A., & Sivakumar, S. (2018). Managerial support for development and turnover intention: Roles of organizational support, work engagement and job satisfaction. *Journal of Organizational Change Management*, *31*(1), 135–153.

Kusluvan, S., Kusluvan, Z., Ilhan, I., & Buyruk, L. (2010). The human dimension: A review of human resources management issues in the tourism and hospitality industry. *Cornell Hospitality Quarterly*, *51*(2), 171–214. doi.org/10.1177/1938965510362871.

Labov, B. (1997). Inspiring employees the easy way. *Incentive*, *171*(10), 114–118.

Li, X., Ma, E., & Qu, H. (2017). Knowledge mapping of hospitality research – A visual analysis using CiteSpace. *International Journal of Hospitality Management*, *60*, 77–93. doi.org/10.1016/j.ijhm.2016.10.006.

Lutz, M. M., & Lux, K. (1988). *Humanistic economics: The new challenge*. New York: The Bootstrap Press.

Maertz, C. P., & Campion, M. A. (2004). Profiles in quitting: Integrating process and content turnover theory. *Academy of Management Journal*, *47*(4), 566–582. doi.org/10.2307/20159602.

Meaghan, S., & Bontis N. (2002). Voluntary turnover: Knowledge management-Friend or foe? *Journal of Intellectual Capital*, *3*(3), 303–322. doi: 10.1108/14691930210435633.

Melè, D. (2003). The challenge of Humanistic Management. *Journal of business Ethics*, *44*, 77–88.

Melè, D. (2016). Understanding Humanistic Management. *Humanistic Management Journal*, *1*(1), 33–55.

Mitchell, T., Holtom, B., Lee, T., Sablynski, C., & Erez, M. (2001). Why people stay: Using job embeddedness to predict voluntary turnover. *Academy of Management Journal*, *44*, 1102–1121. doi.org/10.5465/3069391.

Meyer, J. P., & Allen, N. J. (1997). *Commitment in the workplace*. Thousand Oaks, CA: Sage Publications.

Nieves, J., & Quintana, A. (2018). Human resource practices and innovation in the hotel industry: The mediating role of human capital. *Tourism and Hospitality Research*, *18*(1), 72–83.

Morandin, G., & Bergami, M. (2014). Schema-based sense making of the decision to participate and its effects on job performance. *European Management Review*, *11*(1), 5–20. doi.org/10.1111/emre.12019.

Pirson, M., & Kimakowitz,E. V. (2014). Towards a human-centered theory and practice of the firm. Presenting the humanistic paradigm of business and management. *Journal of Management for Global Sustainability*, *2*(1), 17–48.

Pirson, M., & Lawrence, P. (2010). Humanism in business – Towards a paradigm shift? *Journal of Business Ethics*, *93*(4), 553–565.

Rust, R., Stewart, G., Miller, H., & Pielack, D. (1996). The satisfaction and retention of frontline employees: A customer satisfaction measurement approach. *International Journal of Service Industry Management*, *7*(5), 62–80. doi.org/10.1108/09564239610149966.

Sirgy, M., & Lee, D. (2018). Work-life balance: An integrative review. *Applied Research Quality Life*, *13*, 229–254. doi.org/10.1007/s11482-017-9509-8.

Stamolampros, P., Korfiatis, N., Chalvatzis, K. , & Buhalis, D. (2019). Job satisfaction and employee turnover determinants in high contact services: Insights from employees' online reviews. *Tourism Management*, *75*, 130–147. /doi.org/10.1016/j.tourman.2019.04.030.

Stanley, L., Vandenberghe, C., Vandenberg, R., & Bentein, K. (2013). Commitment profiles and employee turnover. *Journal of Vocational Behavior*, *82*, 176–187. doi.org/10.1016/j.jvb.2013.01.011.

Su, Z. X., Wright, P. M., & Ulrich, M. D. (2018). Going beyond the SHRM paradigm: Examining four approaches to governing employees. *Journal of Management*, *44*(4), 1598–1619.

Tracey, J. B., & Hinkin, T. (2008). Contextual factors and cost profiles associated with employee turnover. *Cornell Hospitality Quarterly*, *49*(1), 12–27. doi.org/10.1177/0010880407310191.

Treuren, G. (2019). Employee embeddedness as a moderator of the relationship between work and family conflict and leaving intention. *The International Journal of Human Resource Management*, *30*(17), 2504–2524. doi: 10.1080/09585192.2017.1326394.

US Bureau of Labour Statistics. (2020). *Annual quits rates by Industry*. Available at: https://www.bls.gov/news.release/jolts.t18.htm.

Van Dick, R., Christ, O., Stellmacher, J., Wagner, U., Ahlswede, O., Grubba, C., Hauptmeier, M., Hohfeld, C., K.Molzen, , & Tissington, P. (2004). Should I stay or should I go? Explaining turnover intentions with organizational identification and job satisfaction. *British Journal of Management*, *15*, 351–360.

Wasmuth, W. J., & Davis, S. W. (1983). Why employees leave. *Cornell Hotel and Restaurant Administration Quarterly*, *24*(1), 11–18.

Weng, Q., & McRoy, J. (2012). Organizational career growth, affective occupational commitment and turnover intentions. *Journal of Vocational Behavior*, *80*, 256–265.

Wright, P. M., & McMahan, G. C. (1992). Theoretical perspectives for strategic human resource management. *Journal of Management*, *18*(2), 295–320. doi.org/10.1177/014920639201800205.

Zimmerman, R., Swider, B., & Boswell, W. (2019). Synthesizing content models of employee turnover. *Human Resource Management*, *58*(1), 99–114. doi.org/10.1002/hrm.21938.

Part II

Innovating Education, Training, and Experiences in Tourism

7 Tourism Education from a Humanistic Perspective

Anthropological Inspirations

Sabina Owsianowska

1 Introduction

The history of anthropological research on contemporary travel dates back to the 1970s. The breakthrough can be considered the Annual Meeting of the American Anthropological Association, which resulted in a volume edited by Valene Smith, *Hosts and Guests. The Anthropology of Tourism* (1977). Tourism became a more recognizable area of research, and anthropologists started to study previously ignored topics related to the expansion of this global industry. Initially, they focused on popular tourist destinations, valued for their exoticism, the possibility of contact with unspoiled nature and authentic indigenous culture, but economically underdeveloped and backward. Long-term fieldwork revealed controversies linked to international mass tourism, and researchers proposed different frameworks to analyze travelers' behaviors, challenges, and pitfalls of intercultural communication, or resistance to the commodification of heritage (Nash, 1996; Selwyn, 1996; Smith, 1977/1989; Smith & Brent, 2001). Currently, the range of topics associated with anthropology is much broader and applies to both traditional and new paradigmatic approaches (Andrews, Jimura & Dixon, 2019; Bloch, 2018; Burns, 2004; Giovine di, 2011; Leite, Castañeda & Adams, 2019; Salazar, 2017; Salazar & Graburn, 2016).

Although the anthropology of tourism is not always considered a separate sub-discipline by anthropologists themselves (Leite & Graburn, 2009), it is the source of many projects at both local and international levels, such as networks, conferences, edited volumes, and thematic issues of scientific journals. Leite and Graburn write about 'anthropological interventions' in interdisciplinary tourism research (Ibid.), while Leite and Swain emphasize that the anthropology of tourism is "consequently defined not by a unified theoretical approach, nor by particular topics, but a common interest in socio-cultural phenomena that transpire in tourism-related settings" (Leite & Swain, 2015, p. 2). This ongoing debate influences strategies of incorporating anthropological ideas, theories, and methods into tourism education.

In this chapter, the place of anthropology in tourism studies is considered on the example of courses offered by Polish universities. Taking into account a broader and global context, two main directions of anthropologically oriented tourism research and education can be highlighted. On the one hand, representatives of ethnology and cultural anthropology, as well as other social sciences and humanities, examine various aspects of travel in the context of their disciplines and focus on theoretical considerations and descriptions of reality. On the other, tourism researchers usually refer to anthropology and sociology, among others, not only to explain this phenomenon but also to apply the results of research (Banaszkiewicz & Owsianowska, 2017, 2018). For tourism scholars, approaches that go beyond traditional long-term field observations are particularly useful because of the ephemeral nature of this type of human mobility (Kaaristo, 2018; Ingold, 2014; Rist, 1980). They often decide on short-term fieldwork that requires repeated returns to the site(s) being examined. Analysis of reality in light of the mobility paradigm allows paying attention to the dynamics of changes in the relationships between guests and hosts, for example, under the influence of global and local processes. As Kaaristo writes, nowadays "the ideas of the natives, tourists, migrants, visitors, etc., become more and more blurred and fluid" (Kaaristo, 2018, p. 72; Selwyn, 2017; Stronza, 2001).

Undoubtedly, anthropological knowledge is crucial for better preparation of specialists in the field of tourism. Managers, hotel employees, tour guides, heritage interpreters, authors of informational and promotional materials, cultural brokers, etc. – all deal with such issues as encounters of tourists with other people, cultural conflicts and misunderstandings, host-guest relationships, or the dissonance of heritage (Hollinshead, 2002; Mostafanezhad & Hannam, 2016; Nash, 1996, 2007; Przecławski, 1997b; Smith, 1977/1989; Smith & Brent, 2001; Tunbridge & Ashworth, 1996). Hollinshead emphasizes that tourism cannot be treated only as (trivial) entertainment or a branch of the industry, and refers to the process of mediating tourist experiences. Tourism staff not only provide information and services (accommodation, transport, guiding, etc.) but above all, they create, promote and share worldviews, values and meanings – they are responsible for a kind of 'world-making' (Hollinshead, 2002, p. 2; Krippendorf, 1987; Lester & Scarles, 2016). In fact, they shape tourists' encounters, practices, and performances, and may contribute to the protection of human dignity, ontological security, traditional lifestyles, and endangered heritages, which is why their knowledge, awareness, sensitivity, and 'soft' skills matter.

Tourism scholars recognize that it is necessary to redefine traditionally understood 'managerial tasks', which treated effectiveness and efficiency as the main objectives (Kostera, 2015; Kimakowitz von, Pirson, Spitzeck, Dierksmeier, & Amann 2011). This can be achieved through education that refers to the people-oriented management paradigm and provides students with adequate knowledge and tools. How anthropology could help in implementing the approach based on three principles of humanistic management: (1) unconditional respect for human dignity; (2) integration of ethical reflection in management decisions; and (3) active and ongoing engagement with stakeholders? (Kostera, 2015; Kimakowitz von et al., 2011) What is and should be examined and taught so that students gain better orientation in the contemporary world and the abilities to deal with the complexity of tourism (Graburn, 2018; Przecławski, 1997b; Selwyn, 2017; Zowisło & Kazimierczak, 2018)? An attempt to find answers to the above and related questions indicates the role of anthropology in tourism education.

In the second part of this chapter, the author refers to the examples of Polish institutions, focusing *per se* on tourism studies, for example, faculties of tourism and leisure, or those offering courses related to tourism and anthropology within other organizational structures and disciplinary frameworks (Alejziak & Winiarski, 2006; Banaszkiewicz & Owsianowska, 2017, 2018). First of all, a course program largely depends on the interest of a lecturer and her/his knowledge of thematic literature, methodologies, etc. However, ideological reasons may influence the choice of themes and case studies or theoretical framework, as well as the avoidance of a 'controversial' field of research (such as gender issues or ethnic and minority communities). The risk of rejection of a program or an application inconsistent with (tacitly) substantive and non-substantive criteria, and as a consequence the lack of funding and institutional support, may therefore determine research and teaching (Banaszkiewicz & Owsianowska, 2017, 2018; Diekmann & Smith, 2015). Last but not the least, from the very beginning of tourism studies, business-based and positivist paradigms dominated. Therefore, initiatives anchored in the economy, marketing, and similar areas of interest, as well as quantitative research methods prevailed over projects examining tourism as a socio-cultural phenomenon, significant human activities, interpersonal communication, and a tool for individual and collective fulfillment (*The Global Code of Ethics for Tourism*).

The humanistic turn in management sciences has shaped new perspectives in teaching staff for tourism and tourism-related industries (Dredge, Airey & Gross, 2014; Cuffy, Airey & Papageorgiou, 2018; Tribe, 1997). The pursuit of sustainable development – understood as a compromise between the expectations and goals of various actors of global and local markets in relation to all beings, natural resources, and cultural diversity – is the basis for remodeling our common future. Nevertheless, sustainability is also linked to citizens' and consumers' behavior, and the creation of lifestyles in accordance with 'sustainable thinking and living' should start at primary school. As Dettori, Caboni, and Giudici state, "the multidisciplinary nature of sustainability also requires new and specific knowledge. This means the necessity of a critical rethinking of the education system in the way to provide the cognitive tools and applications needed for new generations to address environmental, economic, and social challenges" (Dettori, Caboni, Giudici, 2017, p. 65).

2 Anthropology of Tourism, Humanistic Management, and Tourism Studies

What is the relationship between contemporary tourism, tourism anthropology, and humanistic management tenets? In contrast to management theories of the 20th century – the main goals of which are profit, efficiency, and productivity – the principles of humanistic management focus on human needs and values (Kimakowitz von et al., 2011; Kociatkiewicz & Kostera, 2013; Nierenberg, Batko & Sułkowski, 2015; Sułkowski & Woźniak, 2014). The most important is unconditioned respect for the dignity of all persons. In the tourism system (Leiper, 2004; Mill & Morrison, 2012), the complex relationships between tourists (guests), residents (hosts), employees of the tourism sector, state administration, NGOs, private entrepreneurs, etc., should be taken into account. All actions and decisions must be analyzed in light of ethics for all stakeholders, and dialogue is the best way of implementing strategies to achieve a common good (Nierenberg et al., 2015).

The concept of humanistic management is evoked in this chapter in relation to tourism – a multidisciplinary and multidimensional phenomenon, defined by the United Nation World Tourism Organization as follows:

"Tourism comprises the activities of persons traveling to and staying in places outside their usual environment for not more than one consecutive year for leisure, business, and other purposes". (*Glossary of tourism terms,* UNWTO).

This definition has been coined mainly for statistical purposes and does not strictly refer to issues relevant to the anthropological perspective, such as the social and cultural nature of tourism and the behavior of tourists (Przecławski 1997b), including ethical challenges and requirements.

The ethics of tourism is an important area of theoretical reflection and actions, which include such initiatives as the creation of the Global Code of Ethics for Tourism (The Code of Ethics) (Global Code of Ethics for Tourism UNWTO; Fennell, 2006; Przecławski, 1997a, b; Zowisło & Kazimierczak, 2018) and other guidelines, recommendations or certificates of good practice. According to Przecławski, who co-created (with, among others, Marie-Françoise Lanfant) the philosophical foundations of The Code of Ethics, "school, mass media, tour guides, and guidebooks have a great role to play in reminding tourists (and potential tourists) of their duties" (Przecławski, 1997b, p. 25). He emphasizes that they are "responsible for the word, for the truth, and reliability of information" (Ibid, p. 151). Therefore, this requires focusing not only on discourse influencing tourists' imaginaries, expectations, and practices, but also the dissemination of stereotypes, euphemisms, and hyperboles used for marketing purposes (Dann, 1996; Salazar & Graburn, 2016; Owsianowska, 2014).

Dann, the author of *The Language of Tourism. A Sociolinguistic Perspective* (1996), claims that the language of tourism fulfills the function of social control among tourists. In addition, due to its tautological nature, it strengthens stereotypes circulating widely in the public and private space (Dann, 1996, 2011; Jaworski & Thurlow, 2010; Lester & Scarles, 2016; Urry & Larsen, 2011). To identify commonly used clichés, Dann proposes various registers of tourism discourse ('gastrolinguo' for culinary tourism; 'Old'talk' for heritage tourism; 'Spasprech' for health tourism, 'Greenspeak' for ecotourism, etc.). As tourism discourse is essentially multimodal, it dynamically changes in the age of cyberculture (Bielenia-Grajewska & Cortes de Los Rios, 2017; Dann, 2012; Francesconi, 2014; Jaworski & Thurlow, 2010). New communication technologies enrich verbal messages, mainly via visual and sound layers. Social media make it possible to share memories, testimonies, reviews, photos, and videos while traveling, and create a sort of interactive platform for the 'trialogue' between visitors, inhabitants, and tourism staff (Dann, 2012). Visual narratives include selected, idealized, and often retouched images of people and places (Dann, 1996, 2011; Selwyn, 1996). It is noteworthy that representations of residents can lead to their depersonalization or objectification; for example, in advertisements and catalogs, they can only be shown as an element of the attractiveness of a destination or as people serving holiday-makers (Dann, 1996, 2011; Dielemans, 2011; Krippendorf, 1987; Wieczorkiewicz, 2008).

As stated above, the ethical dimension of contemporary travel applies to all participants of tourist interactions. In the second half of the 20th century, criticism of mass tourism and the growing awareness of its negative impact on people, culture, nature, and local economies (e.g., so-called 'leakages'), contributed to the promotion of sustainable tourism development (STD). Sustainable tourism is defined as:

"tourism that takes full account of its current and future economic, social and environmental impacts, addressing the needs of visitors, the industry, the environment, and host communities." (UNEP & UNWTO, 2005, p. 11–12).

Although this idea is strongly supported by scientists and practitioners, its implementation still requires negotiations and solutions that would be responsible and acceptable for all stakeholders (Wood, 2017). Difficulties include, but are not limited to, balancing economic, ecological, socio-cultural benefits and losses in all areas. Minimizing the negative effects of tourism industry development is a difficult task due to the interdisciplinary and multi-dimensional nature of this phenomenon. For example, tourism products should be designed and managed in accordance with the principles of eco-development. Certificates awarded by international, regional, and local organizations distinguish people, places, objects, routes, events, and institutions successfully implementing STD in projects beneficial for the local community and economy, those friendly to nature and culture (see, among others, The One Planet Sustainable Tourism Program – UNWTO; Green Globe Certification – Certified Sustainability; European Destinations of Excellence, EDEN – the European Commission; Certified Tourism Products – the Polish Tourist Organization; Owsianowska & Banaszkiewicz, 2015; Owsianowska, 2014).

Heritage and its management are among key issues in anthropologically inspired tourism research and education. Bearing in mind that tourism itself may not be a source of (dis)advantages, but can play the role of a 'catalyst' for changes in a place, community, local economy, or environment, particular attention should be paid to interpretative strategies. Narratives about culture, history, and values may be distorted, misunderstood, or ideologized (Dann, 1996, 2011; Mikos v. Rohrscheidt, 2018; Owsianowska, 2018; Owsianowska & Banaszkiewicz, 2015). In this approach, heritage is perceived not only as (in)tangible resources (monuments, customs, etc.), but primarily as a relationship between people and the past. According to Tunbridge and Ashworth (1996), dissonance is an intrinsic feature of heritage; therefore, its meaning depends on personal stories and emotional relationships with the past, family, ethnic group or nation, landscapes, etc. The current political, legal, and socio-cultural situation may determine the interpretation of heritage for touristic purposes and beyond (Mikos v. Rohrscheidt, 2018; Owsianowska & Banaszkiewicz, 2015). For example, presenting a nation as homogeneous, regardless of its origin and demographic structure can lead to excessive simplification and perpetuation of stereotypical images. Of particular importance is the official message, supported by the authority of such institutions as national, regional, and local tourist organizations, responsible for creating the Authorized Heritage Discourse (AHD) (Smith, 2006). For this reason, tourism education should be more focused on deepening knowledge on the past, cultural changes, and understanding the complex processes determined by local, national, and global politics.

3 Strategies for Incorporating Anthropological Knowledge into Tourism Education

In this part of the chapter, the main strategies for incorporating topics, contexts, and research methods inspired by anthropology into tourism education are discussed. The author refers to her own professional experience, as well as cooperation with scholars from various institutions in Poland and other countries (i.e., international EU projects; ERASMUS)[1]. For political reasons, the traditions of tourism education in Central and Eastern Europe differ from the experience in Western

Europe or the United States – the so-called Anglophone Snake (Graburn, 2018; Banaszkiewicz, Graburn & Owsianowska, 2017). It is worth mentioning that one of the most characteristic elements of the legacy of the socialist system is the strong link between tourism, recreation, and physical culture/sports, at the same time in practical, scientific, and didactic terms. For this reason, the first faculties of tourism and recreation (e.g., in Poznań and Kraków) were established at universities (previous academies or high schools) of physical education in the 1970s (Banaszkiewicz et al., 2017; Banaszkiewicz & Owsianowska 2017, 2018; Raciborski & Winiarski, 2017). Despite systemic changes after 1989, they still play an important role in interdisciplinary tourism research, the work of experts in the socio-economic environment, and education, including the development of curricula for tourism staff. They have provided templates for universities and faculties or institutes with different profiles (e.g., economics), offering studies in the field of tourism and recreation, tourism management, etc.

Due to the fact that education and research on tourism operate in many fields of knowledge, studies are conducted in such departments as tourism and leisure, physical education and sports, economics, international relations, environmental sciences, cultural studies, philology, management, and even theology (where classes within the 'religious tourism' specialization are proposed). There are also private higher education schools with a touristic profile, for example, High School of Tourism and Foreign Languages or High School of Ecology and Tourism. What is more, the courses called 'anthropology of tourism', 'travel anthropology', 'anthropology of travel and tourism', 'tourism as a cultural experience,' and the like are taught – compulsorily or optionally – as part of various non-tourist studies, for example, ethnology and cultural anthropology, philology (Balkan studies, Italian studies, and others), cultural studies, history, sociology, intercultural and international relations (Banaszkiewicz & Owsianowska, 2017, 2018).

What strategies for incorporating anthropological knowledge into tourism education have been identified? First of all, there are general courses related to socio-cultural anthropology, and tourism is one of the topics of lectures, classes, workshops, and seminars. Secondly, there are courses focused on tourism anthropology, such as 'Anthropology of tourism', 'Anthropology of travel', 'Anthropology of identity (Anthropology of tourism: narratives, identity, place)', 'Cultural anthropology'. The importance of the discipline in the development of 'responsible tourism', etc. Thirdly, there are other courses in the curricula of tourism studies, widely referring to anthropology, for example, 'Cultural tourism', 'Heritage interpretation', 'Intercultural interpretation of dissonant/difficult heritage'. Finally, there are courses from other fields of study, including some anthropologically inspired tourism themes (for instance, the problem of 'thanatourism' as a form of mass tourism and escape from the fear of death as part of the course 'Thanatoeducation in medicine', taught in the field of pedagogy at the Faculty of Health Sciences).

3.1 Tourist Issues in General Courses of Socio-cultural Anthropology

The purpose of the courses is to provide knowledge on the main concepts in the domain of ethnography, ethnology, folklore studies, and socio-cultural anthropology (e.g., people, nation, ethnic group, folk, and national culture, homeland, identity, memory, etc.). Students learn how to use anthropological theories and methods to study contemporary travel. They acquire basic knowledge in the field of anthropology, with particular emphasis on cultural heritage, and are expected to learn about the traditions of local communities in ethnographic regions in Poland and Europe. Other issues include heritage – both natural and cultural, and material and immaterial – as a source of knowledge; social ecosystem and local and regional development; environmental impact; local community; determinants of identity formation; markers of the cultural identity of an individual or a group; the relationships between local, regional, national, and European identities; cultural borderlands; protection of ethnic groups; multiculturalism of Polish and European regions; methods of popularizing knowledge about cultural heritage and its importance in educational and cultural tourism. The context of the historical, legal, and political situation of various ethnic groups living in the region provides information about the creation of cultural landscapes.

In this group, there are, for instance, courses such as 'Ethnography of Poland and Europe' – which is offered to students of tourism and recreation specializing in agritourism at the Faculty of Geography; 'Ethnography and cultural heritage of the region' and 'History of culture' – for students of tourism and recreation at the Faculty of Economics; 'Cultural anthropology' and 'Culture animation' – for students of tourism and recreation at the Faculty of Tourism and Recreation. Tourism themes also appear in courses such as 'Applied anthropology' and 'Anthropology of memory' for students of ethnology and cultural anthropology.

3.2 Tourism Anthropology Courses

In these courses, tourism is in the center of interest of lecturers and students. They discuss basic concepts used in the anthropological analyses of phenomena associated with travel and tourism. On the one hand, students should be able to explain historical and cultural changes affecting travel and tourism practices. They should know the ways in which tourism becomes an inseparable factor in transformation and globalization. On the other hand, students gain knowledge about the impact of tourism research on the reformulation of traditional anthropological concepts. In line with this strategy, they learn about tourism histories and theories; discuss the distinction between tourism and travel, as well as tourism and other forms of mobility (migrations, commuting, etc.). Well-known concepts – such as tourism and tourist typologies and travel modes (Cohen); tourist experience treated as a search for authenticity (MacCannell) or for a new spiritual center (Cohen); the concept of liminality (Turner); 'the tourist gaze' (Urry) and others – are included in the syllabi.

As mass tourism is both an important contribution to the economy of many countries and the cause of various dysfunctions in tourism destinations, students learn how to implement the principles of sustainable tourism development in different sites. Other areas of interest are new forms of travel, such as alternative and niche tourism; cultural and ethnic tourism as an opportunity to encounter cultural diversity; tourism from a postcolonial perspective; visuality and tourism – the role of photography and film; Internet, social media, augmented reality (AR), etc. Students consider, for example, whether the use of smartphones and a widely available Wi-Fi network significantly changes (or not) physical movement as a distinctive type of human activity, that is 'corporeal travel' (Urry 2009) compared to virtual travel. Moreover, the goal of anthropology and sociology of tourism courses is to develop skills in the critical treatment of cultural texts/artifacts, such as travelogues, guidebooks, advertising, and other promotional materials, souvenirs, photographs, and cinematic representations of places, documentaries, etc.

3.3 Anthropological Knowledge Regarding other Courses in the Curricula of Tourism Studies

Courses such as 'Cultural tourism', 'Tourism at heritage sites', or 'Introduction to tourism' are addressed both to Polish and international Erasmus students in the field of tourism and recreation/sports. Travelers interested in other cultures willingly enter the space of the everyday life of others; therefore, the task of universities, faculties, and institutes with a tourist profile and/or specialization should be to promote attitudes of respect towards hosts, their traditions, and different points of view. Tourism staff should be encouraged to develop their empathy, the ability to engage in the roles of people they meet in and outside the workplace. Providing visitors with relevant information, both at tourist information centers, online, and at the facility itself, can be considered the most effective 'tool' for shaping tourist behavior. For instance, "a well-thought-out trip, thoroughly constructed and pursuing educational goals" may contribute to sustainable and responsible tourism (Buczkowska & Malchrowicz-Mośko, 2012, p. 52).

'Basics of tourism and recreation' is a type of introductory course offered to students of tourism at various universities, faculties, and of different specializations. Taking care of health, physical activity, and well-being is an important goal for participants of recreational trips. Vacationers want to keep a work-life balance and maintain a high level of quality of life (Inglehart, 1997). These values have gained particular significance in the post-industrial (postmodern) era in countries of the so-called Global North. The improvement of living conditions in the emerging economies of the less developed regions means that their inhabitants are also beginning to benefit more from tourism, that is they travel internationally more often than in the past when under political and economic constraints (UNWTO Tourism Highlights, 2018; Alejziak, 2018; Cohen, 2011; Owsianowska & Banaszkiewicz, 2018). Therefore, students learn about the motives, values, and cultural patterns of travel rooted in various local, regional, and national traditions. Anthropological and sociological concepts of types of tourists (Cohen) or the stages of the tourist experience (Jafari) frame the basic knowledge on travelers' behavior (Owsianowska, 2019; Podemski, 2005; Przecławski, 1997a; Różycki, 2009).

Another example is the 'Intercultural interpretation of difficult/dissonant heritage', an innovative course for students of cultural studies, specialization of 'Intercultural relations in tourism'. During lectures and workshops, students gain knowledge in the field of tourism and intercultural relations, discuss topics associated with social communication and tourism discourse and acquire skills needed to interpret cultural heritage (cf. Tilden, 2007). They learn how to deal with difficult/dissonant heritage that can cause interpretative problems (e.g., multicultural heritage; heritage commemorating current and past traumatic events, wars, genocide, Holocaust; the ambivalent heritage of the post-socialist and post-industrial era). Knowledge about the risk of stereotyping, simplifications, and possible distortions in presenting facts is crucial in light of the so-called 'post-truth' or 'historical policy'. Critical analysis of (authorized) heritage discourse, identifying different points of view, intentions, and alternative ways of interpretation are the basis for preparing a reliable and responsible message about difficult heritage in intercultural relations.

4 Concluding Thoughts

The dominance of the business approach in tourism research has long limited projects focused on tourism as a cultural phenomenon, while the primacy of the (post)positivist paradigms favors quantitative and statistical methods over qualitative and interpretative methods. However, as the previous Secretary General of UNWTO – Talib Rihai stated in 2012 – tourism is "not only an important economic activity, but millions of conversations and interactions in which visitors and host communities participate every day. And precisely because tourism means exchanging ideas and beliefs, it can be one of the most effective tools for promoting mutual understanding, tolerance and peace" *(Faces of the One Billion,* UNWTO). These words accompanied the social campaign 'One Billion Tourists, One Billion Opportunities', which aimed to make travelers aware that their decisions and actions affect the income of a local community, creating jobs, saving water and energy, protecting natural and cultural heritage, promoting tolerance and respect toward others, etc. ('One Billion Tourists, One Billion Opportunities'.) Tourists themselves could vote for the most responsible travel behaviors and chose the slogans: 'Buy local products', 'Respect local culture', 'Save heritage', 'Save energy', 'Use public transport' as the most important. They also shared their own visual travel narratives in an album 'Faces of the One Billion' on social media (Faces of the One Billion). This UNWTO initiative encouraged the implementation of The Code of Ethics (Handszuh, 2018). Significantly, this document concerns tourism workers and entrepreneurs as well and reminds their fundamental rights, which strictly corresponds to humanistic management that "is oriented not only to obtaining results through people, but also, and above all, toward people themselves, showing care for their flourishing and well-being" (Melé, 2016, p. 33; *The Code of Ethics*, Article 9; Pirson, 2017).

Ethical dilemmas about the further development of tourism have increased dramatically in the 21st century. On the one hand, tourists enjoy the right to travel, to express their freedom of choice, and experience the wealth of the world. On the other, awareness of tourism dysfunctions stimulates criticism. Arguments regarding the well-being of inhabitants of tourist destinations, economic growth or chances of employment and social advancement, etc., do not dispel doubts. A German journalist, Jennie Dielemans, who wrote *Welcome to Paradise. The Reports on the Tourist Industry* admits that she would have probably given up traveling if she had known all the disadvantages of contemporary tourism beforehand (Dielemans, 2011). The manifestos of anti-tourists and counter-tourists sound similar (Sahaj, 2013). In recent decades, climate change, unsustainability, problems with overtourism, and short-term house rental in the most attractive cities and the like have caused changes in (declared) vacation patterns (e.g., promotion of 'staycations' and domestic tourism) or the emergence of new trends in traveling (e.g., 'the shame of flying' or 'slow tourism' movement).

Nevertheless, it is worth looking for ways out of the impasse. First of all, tips on how to travel in a more responsible way are appreciated (de Botton, 2003; Owsianowska, 2014; Zowisło & Kazimierczak, 2018). This applies not only to travel arrangements, but also to philosophical issues: how can tourists obtain knowledge and skills that improve 'the art of travel'? Answers may vary: from the creation of codes of conduct and entrepreneurship by international and local expert bodies, through ethical tourism workshops and networks supporting new modes of travel (Fairbnb; Post-turysta.pl) to writing alternative guides. Education is first and foremost: formal, informal, and non-formal lifelong learning methods targeted at travelers, local communities, and tourism experience mediators (Cuffy et al., 2018; Dredge et al., 2014; Owsianowska, 2018; Tribe, 1997). These tools, aimed at transferring knowledge, shaping attitudes and awareness about one's own culture and that of others, include critical thinking about the role of stereotypes, ethnocentrism, and xenophobia in tourism and beyond (Hollinshead, 2002; Przecławski 1997a), about environmental threats, and ways to minimize the unwanted impact.

The protection of human dignity and the intrinsic value of life as well as concern for the common good are the subjects of reflection of both tourism ethics and humanistic management. In practice, this applies to the care of all persons and groups involved in tourism: visitors, residents, and staff, as well as the welfare of animals and wildlife. An employee's self-awareness in tourism is one of the key competencies because "we all unintentionally decontextualize and recontextualize the world in our work, as we breathe" (Hollinshead, 2002, p. 2). Anthropology is the best source of methods enabling observation and understanding of travelers' behavior and interactions on the one hand, as well as habits of locals, traditions, and everyday life, beliefs, and rituals on the other. Acquiring knowledge about oneself and others, about different cultures and intercultural relations, allows students to look more critically at this inevitable aspect of work in tourism. The responsibility of professionals from the tourism industry, their self-reflexivity, ethical attitudes, and adequate responses to changing local and global realities are as important as managerial know-how and technical skills.

Acknowledgement

The chapter was prepared within the framework of a research project No. 156/BS/KTiR/2018, funded by the Ministry of Science and Higher Education in Poland: Research potential, Chair of Tourism and Leisure, Krakow University of Physical Education.

Note

1 This includes lectures, classes, workshops and seminars conducted since 2006 at (1) the Faculty of Tourism and Leisure, University of Physical Education in Kraków, both for Polish and international students (within the Erasmus Programme); (2) the Institute of Intercultural Studies at the Jagiellonian University in Kraków since 2016, courses: 'Promotion and Information in Tourism' and 'Intercultural Interpretation of Difficult Heritage' (the latter within an innovative specialisation: ReMiT – 'Intercultural Relations in

Tourism', created and directed by M. Banaszkiewicz, Ph.D., anthropologist); (3) the College of Administration in Bielsko-Biała, Poland, for students in International Relations, specialisation of International Tourism, 2008–2014 and two EU projects between Polish, Czech and Slovakian institutions, co-ordinated by Mrs. B. Górka; (4) the Faculty of Tourism, University of Primorska, Koper, Slovenia – visiting professor in 2009 and lecturer during summer school in 2010 (Intensive Programme, Erasmus), courses and workshop centred on intangible heritage in tourism.

References

Alejziak, W., (2018). Research on Tourist Activity in Europe – Methodological Aspects and Results (Part I: the Period 1945–2000). *Folia Turistica*, *46*, 9–48.

Alejziak, W. & Winiarski, R., (2006). *Tourism in scientific research in Poland and worldwide*. Rzeszow & Krakow: WSiZ & AWF Press.

Andrews, H., Jimura, T. & Dixon, L., (2019). *Tourism Ethnographies: Ethics, Methods, Application and Reflexivity*. London: Routledge.

Banaszkiewicz, M. & Owsianowska, S., (2017). Antropologia a edukacja i badania w zakresie turystyki [Anthropology and tourism education and research]. In S. Owsianowska & R. Winiarski (Eds.), *Antropologia turystyki [Anthropology of Tourism]* (pp. 101–112). Kraków: AWF Press.

Banaszkiewicz, M. & Owsianowska, S., (2018). Introduction: Anthropological Studies on Tourism in Central and Eastern Europe. In S. Owsianowska & M. Banaszkiewicz (Eds.), *Anthropology of Tourism in Central and Eastern Europe: Bridging Worlds* (pp. 1–24). Lanham: Rowman & Littlefield.

Banaszkiewicz, M., Graburn, N. & Owsianowska, S., (2017). Tourism in (Post) socialist Eastern Europe. *Journal of Tourism and Cultural Change*, *15*(2), 109–121.

Bielenia-Grajewska, M. & Cortes de Los Rios, E. (Eds.) (2017). *Innovative perspective on tourism discourse*. London: IGI Global.

Bloch, N., (2018). *Bliscy nieznajomi. Turystyka i przezwyciężanie podporządkowania w postkolonialnych Indiach [Close Strangers. Tourism and Overcoming Subordination in Post-colonial India]*. Poznań: UAM Press.

Botton de, A., (2003). *The Art of Travel*. London: Penguin Books.

Buczkowska, K. & Malchrowicz-Mosko, E., (2012). Etyczne dylematy turystyki kulturowej [Ethical dilemmas of cultural tourism]. *Turystyka Kulturowa*, *12*, 42–55.

Burns, G. L., (2004). Anthropology and Tourism: Past Contributions and Future Theoretical Challenges. *Anthropological Forum*, *14*(1), 5–22.

Cohen, E., (2011). The Changing Faces of Contemporary Tourism. *Folia Turistica*, *25*, 13–19.

Cuffy, V., Airey, D. & Papageorgiou, G., (2018). *Lifelong Learning for Tourism. Concepts, Policy, Implementation*. London: Routledge.

Dann, G., (1996). *The Language of Tourism. A Sociolinguistic Perspective*. Wallingford: CAB International.

Dann, G., (2011). 'Take me to the Hilton': A Language of Tourism Paradigm. *Folia Turistica*, *25*, 23–40.

Dann, G., (2012). Remodelling a changing language of tourism: from monologue to dialogue and trialogue. *Pasos. Revista de Turismo y Patrimonio Cultural, 10*(4), 59–70.

Dettori, A., Caboni, F. & Giudici, E., (2017). Education and Sustainability: The Case of Emotions Park. *Journal of Management and Sustainability*, 7, 65–75.

Diekmann, A. & Smith, M. K., (2015). *Ethnic and minority cultures as tourist attractions*. Bristol: Channel View Publications.

Dielemans, J., (2011). *Witajcie w raju. Reportaże o przemyśle turystycznym [Welcome to Paradise. Reports on the tourist industry]*. Wołowiec: Czarne.

Dredge, D., Airey, D. & Gross, M. J. (Eds.) (2014). *The Routledge Handbook of Tourism and Hospitality Education*. London: Routledge.

Fennell, D., (2006). *Tourism Ethics*. Bristol: Channel View Publications.

Francesconi, S., (2014). *Reading Tourism Texts. A Multimodal Analysis*. Bristol: Channel View Publications.

Giovine di, M., (15 April 2011). Tourism Research as "Global Ethnography". *Anthropologies*. http://www.anthropologiesproject.org/2011/04/tourism-research-as-global-ethnography.html.

Graburn, N., (2018). Inside and Outside the Anglophone Snake: Alterities and Opportunities. In S. Owsianowska & M. Banaszkiewicz (Eds.), *Anthropology of Tourism in Central and Eastern Europe: Bridging Worlds* (pp. 27–48). Lanham: Rowman & Littlefield.

Handszuh, H., (2018) Aspekty koncepcyjne i prawno-międzynarodowe Światowego Kodeksu Etyki dla Turystyki [The conceptual and international law aspects of the Global Code of Ethics for Tourism]. *Folia Turistica, 49*, 151–198.

Hollinshead, K., (2002). *Tourism and the making of the world. The dynamics of our contemporary tribal lives*. Miami: Florida International University, The Year 2002 – Honors Excellence Lecture, The Honors College.

Inglehart, R., (1997). *Modernization and Postmodernization. Culture, Economic and Political Change in 43 Societies*. New Jersey: Princeton University Press.

Ingold, T., (2014). That's enough about ethnography. *HAU. Journal of Ethnographic Theory*, 4, 383–395.

Jaworski, A. & Thurlow, C., (2010). *Tourism Discourse: Language and Global Mobility*. Basingstoke & New York: Palgrave MacMillan.

Kaaristo, M., (2018). Engaging with the hosts and guests: Some methodological reflections on the anthropology of tourism. In S. Owsianowska & M. Banaszkiewicz (Eds.), *Anthropology of Tourism in Central and Eastern Europe. Bridging Worlds* (pp. 71–88). Lanham: Rowman & Littlefield.

Kimakowitz von, E., Pirson, M., Spitzeck, H., Dierksmeier, C. & Amann, W., (2011). *Humanistic Management in Practice*. London: Palgrave Macmillan.

Kociatkiewicz, J. & Kostera, M., (2013). Zarządzanie humanistyczne: zarys programu [Humanistic Management: basics of the programme]. *Problemy Zarządzania, 11*(4), 9–19.

Kostera, M., (2015). Refleksje o zarządzaniu humanistycznym [Reflections on humanistic management]. In B. Nierenberg, R. Batko & Ł. Sułkowski (Eds.), *Zarządzanie humanistyczne [Humanistic Management]* (pp. 53–61). Kraków: WUJ.

Krippendorf, J., (1987). *The Holiday Makers. Understanding the Impact of Leisure and Travel*. London: Heinemann.

Leiper, A., (2004). *Tourism management*. Sydney: Person Hospitality Press.

Leite, N. & Graburn, N., (2009). Anthropological interventions in tourism studies. In M. Robinson & T. Jamal (Eds.) *Handbook of Tourism Studies* (pp. 35–64). London: Sage.

Leite, N. & Swain, M., (2015). Anthropology of Tourism. In J. Jafari & H. Xiao (Eds.), *Encyclopedia of Tourism* (2nd ed.). London: Springer. https://eprints.soas.ac.uk/19859/1/Leite_Swain_EncyclTourism.pdf. Accessed July 2019.

Leite, N., Castañeda, C. & Adams, K. (Eds.) (2019). *Tourism Ethnography: Edward Bruner and Beyond*. Lanham: Rowman & Littlefield.

Lester, J. & Scarles, C., (2016). *Mediating the Tourist Experience: From Brochure to Virtual Encounters*. London: Routledge.

MacCannell, D., (2002). *Turysta. Nowa teoria klasy próżniaczej [The tourist. A new theory of the leisure class]*. Warszawa: Muza.

Melé, D., (2016). Understanding Humanistic Management. *Humanistic Management Journal*, *1*, 33–55.

Mikos v. Rohrscheidt, A., (2018). Polityka historyczna z etyką w tle. Patologie w kreowaniu doświadczeń turystów kulturowych [Historical Policy with Ethics in the Background. Pathologies in Creating Experiences of Cultural Tourists]. *Folia Turistica*, *49*, 67–124.

Mill, R. & Morrison, A., (2012). *The tourism system*. Dubuque: Kendall Hunt Publ. Company.

Mostafanezhad, M. & Hannam, K., (2016). *Moral Encounters in Tourism*. London: Routledge.

Nash, D., (1996). *The Anthropology of Tourism*. Oxford: Pergamon.

Nash, D. (Ed.) (2007). *The Study of Tourism. Sociological and Anthropological Beginnings*. Oxford: Elsevier.

Nierenberg, B., Batko, R. & Sułkowski, Ł. (Eds.) (2015). *Zarządzanie humanistyczne [Humanistic Management]*. Kraków: WUJ.

Owsianowska, S., (2014). Wzorce odpowiedzialnego podróżowania w narracjach turystyki kulturowej [Patterns of responsible travelling in cultural tourism narratives]. In M. Kazimierczak (Ed.), *Etyczny wymiar podróży kulturowych [The Ethical Dimension of Cultural Travels]* (pp. 53–66). Poznań: AWF Press.

Owsianowska, S., (2018). Multiculturality and lifelong learning. A Polish case. In V. Cuffy, D. Airey & G. Papageorgiou (Eds.), *Lifelong learning for tourism. Concepts, policy, and implementation* (pp. 127–143). London: Routledge.

Owsianowska, S., (2019). Turystyka w perspektywie antropologii: badania, edukacja, praktyka [Tourism from the perspective of anthropology: research, education, and practice]. In M. Zowisło & J. Kosiewicz (Eds.), *Sport i turystyka w perspektywie nauk spolecznych: tradycja i wsp¢lczesnosc [Sport and tourism from the perspective of social sciences: tradition and contemporariness]* (pp. 278–286). Krakow: AWF Press.

Owsianowska, S. & Banaszkiewicz, M., (2015). Dissonant heritage in tourism promotion. Certified tourism products in Poland. *Folia Turistica*, *37*, 145–166.

Owsianowska, S. & Winiarski, R. (Eds.) (2017). *Antropologia turystyki. Wydawnictwo jubileuszowe z okazji 40-lecia Wydziału Turystyki i Rekreacji AWF w Krakowie [Anthropology of Tourism. Jubilee Edition on the Occasion of the 40thAnniversary of the Faculty of Tourism and Leisure at the University of Physical Education in Krakow]*. Kraków: AWF Press.

Owsianowska, S. & Banaszkiewicz, M. (Eds.) (2018). *Anthropology of Tourism in Central and Eastern Europe: Bridging Worlds*. Lanham: Rowman & Littlefield.

Pirson, M., (2017). *A Humanistic Perspective for Management Research: Protecting Dignity and Promoting Well Being. Humanistic Management Association, Research Paper Series No. 17-8*, 1–32.

Podemski, K., (2005). *Socjologia podróży [The Sociology of Travel]*. Poznań: UAM Press.

Przecławski, K., (1997a). *Etyczne podstawy turystyki [Ethical foundations of tourism]*. Kraków: Albis.

Przecławski, K., (1997b). *Człowiek a turystyka. Zarys socjologii turystyki [Human being and tourism. The outline of tourism sociology]*. Kraków: Albis.

Różycki, P., (2009). *Zarys wiedzy o turystyce [Basics of tourism]*. Kraków: Proksenia.

Raciborski, J. & Winiarski, R., (2017). Organizacja i funkcjonowanie Wydzialu Turystyki i Rekreacji krakowskiej Akademii Wychowania Fizycznego [Organization and activities of the Faculty of Tourism and Leisure at Krakow AWF]. In S. Owsianowska & R. Winiarski (Eds.), *Antropologia turystyki [Anthropology of Tourism]* (pp. 39–52). Krakow: AWF Press.

Sahaj, T. (2013). Journeys Deep Into Oneself and the World. Philosophical Reflections. *Folia Turistica*, *28*(1), 27–34.

Salazar, N., (2017). Anthropologies of Tourism: What's in a Name? *American Anthropologist*, *119*(4), 723–725. https://anthrosource.onlinelibrary.wiley.com/doi/pdf/10.1111/aman.12954.

Salazar, N. & Graburn, N. (Eds.) (2016) *Tourism Imaginaries. A Conceptual Approach*. London: Berghan Books.

Selwyn, T. (Ed.) (1996) *The Tourist Image: Myths and Mythmaking in Tourism*. London: Routledge.

Selwyn, T., (2017) Antropologia podróży, turystyki i pielgrzymowania [Anthropology of Travel, Tourism and Pilgrimage]. In Owsianowska, S. & Winiarski, R. (Eds.), *Antropologia turystyki [Anthropology of Tourism]* (pp. 76–100). Kraków: AWF Press.

Smith, L., (2006). *Uses of Heritage*. London & New York: Routledge.

Smith, V. (Ed.) (1977/1989) *Hosts and Guests. The Anthropology of Tourism*. Pennsylvania: University of Pennsylvania Press.

Smith, V. & Brent, M. (Eds.) (2001). *Hosts and Guests Revisited. Tourism Issues of the 21st Century*. New York: Cognizant Communication Corporation.

Stronza, A., (2001). Anthropology of Tourism: Forging New Ground for Ecotourism and Other Alternatives, *Annual Review of Anthropology*, *30*, 261–283.

Sułkowski, Ł & Woźniak, A. (Eds.) (2014). Zarządzanie humanistyczne [Humanistic Management] (part III. Łódź-Warszawa: Wydawnictwo Społecznej Alademii Nauk). *Przedsiębiorczość i Zarządzanie*, *XV*(11), 1–157.

Tilden, F., (2007). *Interpreting Our Heritage*. Chapel Hill: University of North Carolina Press:.

Tribe, J., (1997). The Indiscipline of Tourism. *Annals of Tourism Research*, *24*, 638–657.

Tunbridge, J. & Ashworth, G., (1996). *Dissonant heritage: the management of the past as a resource in conflict*. Chichester: Wiley.
Urry, J., (2009). *Socjologia mobilności [Sociology beyond societies. Mobilities for the twenty-first century]*. Warszawa: Wydawnictwo Naukowe PWN.
Urry, J. & Larsen, J., (2011). *The Tourist Gaze 3.0*. London: Sage.
Wieczorkiewicz, A., (2008). *Apetyt turysty. O doświadczaniu świata w podróży [The Appetite of a Tourist. On Experiencing the World in Travel]*. Kraków: Universitas.
Wood, M. E., (2017). *Sustainable Tourism on a Finite Planet: Environmental, Business and Policy Solutions*. London & New York: Routledge.
Zowisło, M. & Kazimierczak, M. (Eds.) (2018), *Etyka w turystyce [The Ethics of Tourism]. Folia Turistica 49* [Special Issue].

Internet sources

EDEN – Destinations of Excellence, https://ec.europa.eu/easme/en/cos-teden-2015-3-05-european-destinations-excellence-promotion-eden-destinations-and-awareness, accessed 21 July 2019.
Faces of the One Billion UNWTO, #TravelEnjoyRespect,https://www.facebook.com/pg/ WorldTourismOrganization/photos?tab=album&album_id=287356294716365, accessed 27 July 2019.
GreenGlobe, http://greenglobe.com/, accessed 21 July 2019.
Global Code of Ethics for Tourism UNWTO, http://ethics.unwto.org/content/global-code-ethics-tourism, accessed 21 July 2019.
Glossary of tourism terms, UNWTO, https://www.unwto.org/glossary-tourism-terms, accessed 27 July 2019.
Post-turysta.pl, http://postturysta.pl, accessed 27 July 2019.
The Code of Conduct for the Protection of Children from Sexual Exploitation in Travel and Tourism (the Code), http://www.thecode.org, accessed 27 July 2019.
One billion tourists, one billion opportunities, UNWTO; http://1billiontourists.unwto.org/, accessed 27 July 2019.
UNEP & UNWTO. (2005) *Making Tourism More Sustainable - A Guide for Policy Makers*, http://sdt.unwto.org/content/about-us-5, accessed 27 July 2019.
UNWTO Tourism Highlights 2018 Edition, https://www.e-unwto.org/doi/book/10.18111/9789284419876, accessed 27 July 2019.x.

8 Humanistic Management Training

An Explorative Study on Erasmus Students' Intercultural Sensitivity

Anna Irimiás, Mariangela Franch, and Ariel Mitev

1 Introduction

In 2017, I attended the wedding of a former student of mine. The couple's love relationship started 10 years before, thanks to the Erasmus mobility in Turkey where the Hungarian bride met her Italian groom. The following years required a timely schedule of organization, because of their traineeships and master courses in different European countries. After graduation, they moved to Germany to work for transnational companies. Their intercultural competencies, familiarity with different cultural costumes, and their respect for human diversity were determinant in their successful job application. Many colleagues and friends came to celebrate at their wedding, where Hungarian, Italian, Spanish, Turkish, Dutch, German, and many more languages could be overheard.

This personal story is far to be exceptional. According to the European Commission, one in five Erasmus students in a relationship met their partner while on Erasmus mobility (**E**u**R**opean Community **A**ction **S**cheme for the **M**obility of **U**niversity **S**tudents). The Erasmus exchange program, since its establishment in 1987, has led more than 3 million students to undertake learning periods abroad (ec.europa.eu), which is a relevant tourism mobility *per se*. The annual Erasmus Impact Study tracks students' improvements in diverse skills (Erasmus Higher Education Impact Study, 2019). The acknowledgment of cultural differences, the improvement of interpersonal competencies and communication skills, the increase of social and cultural openness along with a firm standing against discrimination and xenophobia were among the skills, values, and moral principles that students recalled immediately after the Erasmus experience (Erasmus Higher Education Impact Study, 2019:3). There is less evidence, however, on whether behavioral and attitudinal changes were long-term (Rexeisen et al., 2008). In addition, Erasmus mobility provides not only opportunities but challenges as well. Difficulties of financial (additional costs of living abroad) and administrative nature (lack of mutual recognition of exams or complications in transferring earned credits) might delay students' graduation (Shaftel, Shaftel, & Ahluwalia, 2007). Still, it is not surprising that higher education institutions consider Erasmus and Erasmus+ programs and traineeships to be the paramount way to develop intercultural sensitivity and cross-cultural communication skills (Ozcelik & Zoltay Paprika, 2010).

It is assumed that exposure to different cultures automatically contributes to skill development (Farmaki, 2018). Cognitive effort, willingness, and time are required to develop intercultural sensitivity and to enhance cross-cultural communication competencies. The benefits of international experiences are acknowledged when individuals reflect on their experiences and are willing to translate their intentions into attitudinal and behavioral change (Olson & Kroeger, 2001). Thus, reflective individuals are able to effectively interact with people from different cultural backgrounds. Soft skills, along with ethical reasoning, should be at the base of responsible decision-making in different disciplines and fields, such as tourism and management (Pirson, 2019). This explorative study, based on data from management and tourism students, investigates how those who participated in Erasmus mobility evaluate their skills and competencies and studies the role that an Erasmus experience can play in the augmentation of intercultural sensitivity and communication skills. The research questions that drove our study were the following:

RQ_1 How do students self-evaluate their own intercultural sensitivity and communication skills after their Erasmus experience?

RQ_2 How do students evaluate the support of their higher education institutions?

RQ_3 What role can universities play to stimulate students' reflection on intercultural sensitivity?

2 Literature Review

2.1 Importance of Understanding the Role of Intercultural Sensitivity in Cross-cultural Communication

All interactions can be viewed as 'intercultural'. The degree of how an intercultural encounter can be perceived depends on the differences of beliefs, behaviors, attitudes, and social norms between individuals (Reisinger & Turner, 2011). Intercultural sensitivity can be defined as an emotional ability of a person to acknowledge, accept, and behave according to cultural differences (Yurur, Koc, Taskin, & Boz, 2018; Hammer, Bennett, & Wiseman, 2003). In communication, the sender of a message must consider with sensitivity who the receiver is to code the meaning of the message, and the receiver should be able to decode it. The term 'sensitivity' refers to an individual's degree of awareness of changes, internal or external, and to the intrapersonal responsiveness to such changes.

Interculturalist sociologists like Harry Triandis (2006) define 'culture' by distinguishing objective and subjective culture. Objective culture is indicated by a nation's institutions, political, and economic systems (Reisinger & Turner, 2011). Subjective culture is defined by people's worldview within a society. It is shaped by the patterns of attitudes, beliefs, norms, and values (Triandis, 1994). The comprehension of subjective culture provides access to acknowledge a wide spread of worldviews. Intercultural encounters are social interactions between people who are culturally different (Bennett & Bennett, 2004). In our intrinsically networked global era, such interactions occur in the business environment and in everyday life. From the theoretical point of view, there is an ongoing discussion on whether subjective cultures are converging towards commonality or preserve unique values and set of norms. The convergence theory of culture holds that the global flows of people due to immigration, cross-cultural marriages, and travel and tourism intersect subjective cultures to homogeneity (Reisinger & Crotts, 2010). The unprecedented technological advancement eases the diffusion of ideas, brands, fashion, films, and TV series across borders and contributes to making people similar in their needs, desires, and consumption. In contrast, the divergence theory claims that societies are resistant to change worldviews, beliefs, and value systems rooted in history and such worldviews are maintained across regions and countries (de Mooji, 2019).

To explore the extent to which Erasmus students perceive themselves as interculturally sensitive individuals, Milton Bennett's (1986) conceptualization and Developmental Model of Intercultural Sensitivity (DMIS) was deemed appropriate as a theoretical starting point because it is widely used in tourism research (Kirillova, Lehto, & Cai, 2015; Reisinger & Turner, 1998). Rather than a description of attitude and behavior, the model is a cognitive structure of development in six stages. It is based on the perception and behavior of individuals when exposed to interactions with culturally different people (Table 8.1). In intercultural encounters, individuals seek to interpret their experiences using their habitual set of norms and systems of meanings. The model is a continuum showing how getting familiar with different worldviews and understanding a culture as an insider increases competencies in intercultural relations.

The first three stages are defined as ethnocentric. According to Triandis (2006, p. 22) it as "an inescapable reality is that all humans are ethnocentric". Individuals assume that what they consider 'normal' in their culture is 'normal' anywhere. Encountering a different set of norms might be perceived as something 'abnormal' and, consequently, threatening, and risky (Antal & Friedman, 2008). Further, cultural generalizations and stereotyping tend to minimize cultural differences. Stereotyping is perceived as a negative attitude toward another group of people, while

Table 8.1 Developmental model of intercultural sensitivity (dmis) stages

Denial Defence Minimization	}	Ethnocentric stages when one's culture is perceived as central to reality. To be ethnocentric means that you make life choices and act based on the assumption that your worldview is superior.
Acceptance Adaptation Integration	}	Ethno-relative stages when one's culture is perceived in the context of other cultures. Ethno-relativism assumes that behavior can only be understood within a cultural context.

Source: Authors' own elaboration based on Bennett (1986), Hammer (2011) and Hammer, Bennett and Wiseman (2003).

minimization refers to one's own worldview as universal. The second three stages are referred to as ethnorelative. One's own culture is understood within the realm of other cultural worldviews considered equally valid. Being ethnorelative allows people to empathize with others and to understand different sets of norms as insiders. In the stage of Integration, an individual is able not only to comprehend and appraise cultural differences but also to put the self in different worldviews. Consequently, it is assumed that worldviews are not stable but are influenced and shaped by experiences and learning (Peng, Van Dyne, & Oh, 2015).

2.2 *Intercultural Sensitivity and Tourism*

Tourism has turned out to be a global phenomenon, reaching high peak numbers in international tourist arrivals all around the world and especially in Europe (UNWTO International Tourism Highlights, 2019). The ability to understand and appreciate cultural differences and to translate that understanding into effective culturally sensitive communication determines successful tourism marketing and management (Reisinger & Turner, 2011). Intercultural sensitivity is highly valuable in culturally interrelated managerial environments and especially in tourism where intercultural encounters are the essence (Kenesei & Stier, 2016). Traveling and studying abroad modulates the self-image and develops a wide range of self-understandings (a student-identity, tourist-identity, foreigner-identity, etc.) leading to cultural self-awareness (Bennett & Bennett, 2004). Understanding and appreciating others, especially people with a diverse cultural background, is reflected by the understanding of the self (Earley & Ang, 2003). The role played by intercultural sensitivity and communication skills in the tourism industry is outlined in the next paragraphs.

Reisinger & Turner (2011) urged the need to adopt intercultural sensitivity perspectives in tourism business operations. The authors put it clear that 'national ways of thinking can lead to dangerous assumptions about the foreign tourist markets the industry tries to attract such as ethnocentrism, prejudice, and stereotyping' (Reisinger & Turner, 1997, p. 146). In tourism research, some sporadic studies explored the ways intercultural sensitivity influenced interactions between different dyads: tourists-tourists, tourists-locals, and tourists-service providers. Kirillova et al. (2015) applied the Intercultural Sensitivity Scale to explore volunteer tourists' experiences with local people. Reisinger & Turner (1998) studied six dimensions of cross-cultural interaction between hosts and tourists finding that self-actualization, hosts' responsiveness and courtesy, hosts' ability to understand tourists' needs, social interaction, and social obligation are the dimensions where significant cultural differences emerge. The authors called for new service training programs with a focus on intercultural sensitivity. Barron and Dasli (2010) studied how home and international students enrolled in tourism and hospitality courses develop intercultural sensitivity towards one another. Sizoo et al. (2003) measured the effect of intercultural sensitivity on the performance of hospitality employees in cross-cultural service encounters, evidencing that high intercultural sensitivity leads to higher job and social satisfaction.

2.3 Intercultural Sensitivity and Study-Abroad Programs

Intercultural sensitivity is a cognitive process (Jackson, 2015). Past research showed that intercultural sensitivity can be strengthened by properly designed education and training sessions on campus or in study-abroad programs (Bennett & Bennett, 2004). Learning foreign languages is one of these training sessions, and it usually provides insights into the subjective culture of the people and contributes to delineate accurate cultural generalizations (Bennett and Bennett, 2004). Study-abroad programs might be a path to develop intercultural competence, and past studies show that understanding cultural differences might be stimulating for students to become empathetic, socially responsible, and ethical (Nguyen, 2017). This is a key competence for tourism and management students, who, once working in the sector, are going to face several challenging situations where intercultural sensitivity is needed. Study-abroad programs vary in category, length, and credit-hour requirement. Patterson (2006) investigated study-abroad students' intercultural sensitivity compared to on-campus students in the United States. Her findings revealed that students in the classroom began in a similar ethnocentric stage on the *DMIS* (Bennett, 1986) as those students who chose to participate in a study abroad program, but those who studied abroad showed more intercultural development than on-campus students (Patterson, 2006). Similar results were found by Anderson,

Lawton, Rexeisen, & Hubbard (2006). The authors evidenced that short-term, non-language based study programs had a positive impact on the intercultural sensitivity of students, who, in only four weeks, developed critical consciousness about issues of cultural diversity and social justice (Anderson et al. 2006). These studies explored, in part, how study-abroad programs provided students the possibility to widen their worldview and to be able to question their prevailing values and norms.

International study programs such as Erasmus are considered the most appropriate to develop intercultural sensitivity and competences (Peng, Van Dyne, & Oh, 2015). However, the experience of Erasmus study-abroad programs often fails to enhance students' intercultural sensitivity when these are not appropriately supported by the sending institutions and are not designed with a follow-up course stimulating reflection on such experiences (Jackson, 2015).

2.4 *Intercultural Sensitivity and Humanistic Management*

Pirson (2017, p. 237) argued that managerial education needs to be more human-centered, with issues related to 'values, character, trust, empathy, and mindfulness'. Decision-taking and actions led by empathy aim to ensure dignity to others, to embrace social justice issues, and foster cultural diversity. This approach contributes to train socially responsible leaders who embrace individual and social well-being (Pirson, 2018). Diversity management – the ability to recognize, appreciate, and manage cultural differences – is purposeful to efficiently exchange ideas not only among corporate members but in any platform of communication (Westrick, 2004). Reflection is needed to locate one's own habits, behavior, and emotional manifestations into one's own social environment (Hogg & Terry, 2000). In South Africa, Jackson and Haines (2007) studied the multicultural environment of five representative local NGOs to explore the cross-cultural dynamics in the management of people. The results revealed that misunderstandings mainly derived from the different views of the value of people according to the 'modern' Western and to the indigenous humanistic African approaches to value people. The comprehension that we behave according to cultural constructs is far to be natural; rather, it is a cognitive process (Irimiás & Franch, 2019). The dynamics of such a process are influenced by subjective experiences and the perception of such experiences. Giudici et al. (2011) explored the importance of ethical behavior and social commitment in management, focusing on the role universities should play in nurturing and diffusing such values by designing appropriate training sessions and courses. Intercultural sensitivity could be a pathway to such humanistic management education.

We contribute to the existing literature by exploring (1) Erasmus students' experience and their perception of intercultural sensitivity, (2)

Students' evaluation on the support of their higher education institution, in order to question (3) What role can universities play to stimulate students' reflection on intercultural sensitivity? We argue that theoretically grounded and evidence-based quantitative studies on whether and how Erasmus study exchanges contribute to developing students' intercultural sensitivity are underrepresented (Lugosi & Jameson, 2017; Yurur, Koc, Taskin, & Boz, 2018).

3 Methodology

Research instrument: A structured online questionnaire survey was used as the research instrument. In the first part, we attempted to assess participants' mental-health status to capture their current feelings about their Erasmus experience. A list of 16 adjectives was selected from the Affectometer 2 developed by National Health Scotland (2006). Respondents were asked to rate on a 5-point Likert-scale (1 = not at all, 5 = all of the time) how often that items applied to them in the study-abroad period.

In the second part, we adopted Olson and Kroeger's (2001) Intercultural Sensitivity Scale and Intercultural Communication Scale. The research attempted to assess the perception and self-evaluation of students; consequently, the original scale items were reduced and were adopted to reflect Erasmus students' viewpoints. For example, Olson and Kroeger's (2001) scale-item 3 on 'denial', "I feel most comfortable living and working in a community where people look and act like me" was reworded as "I feel most comfortable with people who look and act like me". All 24 scale-items were measured by a 7-point Likert scale (1 = does not describe me at all, 7 = describes me extremely well).

In the third part of the survey, socio-demographic questions were set.

Sample: For this retrospective study, outgoing and incoming Erasmus students enrolled in the academic year of 2018/19 at the University of Trento (Italy) and at the Corvinus University of Budapest (Hungary) were invited to participate. This convenience sample choice of both outgoing (home) and incoming (international) students is motivated by the wish to observe the intercultural sensitivity of management students with different cultural backgrounds (Western-European and Central-European). Only students who spent at least three months abroad were contacted to fill in the survey (Table 8.2).

Data collection: Data collection was carried out from May to June, 2019. The questionnaire was administered once students had returned from Erasmus mobility. In order to protect students' privacy, the Erasmus mobility offices of the respective universities emailed former Erasmus students a link with the online survey powered by Google. Of the 224 completed questionnaires, five were excluded because of duplication, and a total of 219 questionnaires were statistically analyzed.

A total of 219 students responded to the survey. Among the

participants, 31.5% were male and 68.5% female. The sample frame for this survey consisted of students aged between 20 and 28 years, but the age group of 22–24 with 60.3% was overrepresented. Since our focus was on how students evaluate their own intercultural sensitivity and communication skills (RQ_1), it was deemed relevant to investigate whether Erasmus students are from a multicultural family background and are used to switching between different worldviews. As Table 8.3 shows, only about 7% of the participants had a multicultural background.

Table 8.2 The amount of time students spent on Erasmus

Erasmus time					
		Frequency	*Percent*	*Valid Percent*	*Cumulative Percent*
Valid	3–5 months	135	61.6	61.6	61.6
	6–8 months	43	19.6	19.6	81.3
	9+ months	41	18.7	18.7	100.0
	Total	219	100.0	100.0	

Source: Authors' own elaboration.

Table 8.3 Participants' cultural family background

Are your parents from different cultures?					
		Frequency	*Percent*	*Valid Percent*	*Cumulative Percent*
Valid	No	202	92.2	92.2	92.2
	I grew up in a multicultural family	12	5.5	5.5	97.7
	My parents belong to the same culture but they settled down in another country where we live	3	1.4	1.4	99.1
	Prefer not to answer	2	.9	.9	100.0
	Total	219	100.0	100.0	

Source: Authors' own elaboration.

4 Findings

The survey started with an open-ended sentence on the hypothesis of what would have happened if they had not participated in Erasmus mobility. The instrument of the open-ended sentence investigates in an indirect way the perceived outcomes of Erasmus mobility (Figure 8.1). The top-of-mind answers were coded. Table 8.4 shows that participants would have regretted the most (21.9%) the lost 'opportunities', although it is not clear what such 'opportunities' meant for them. 19.6% said that the Erasmus experience significantly contributed to the development of their personality, and it was one of the most relevant experience of their lives, as shown by the next quotes:

> "wouldn't have experienced some of the best moments of my life so far and I wouldn't have met so many amazing people" (a female from Italy)

> "I would have missed a great opportunity. An opportunity to grow up, to meet people with different backgrounds and be able to interact with them. Personally, it helped me to get to know myself better and to get out my comfort zone, improving my ability to stay open-minded." (a female from Germany)

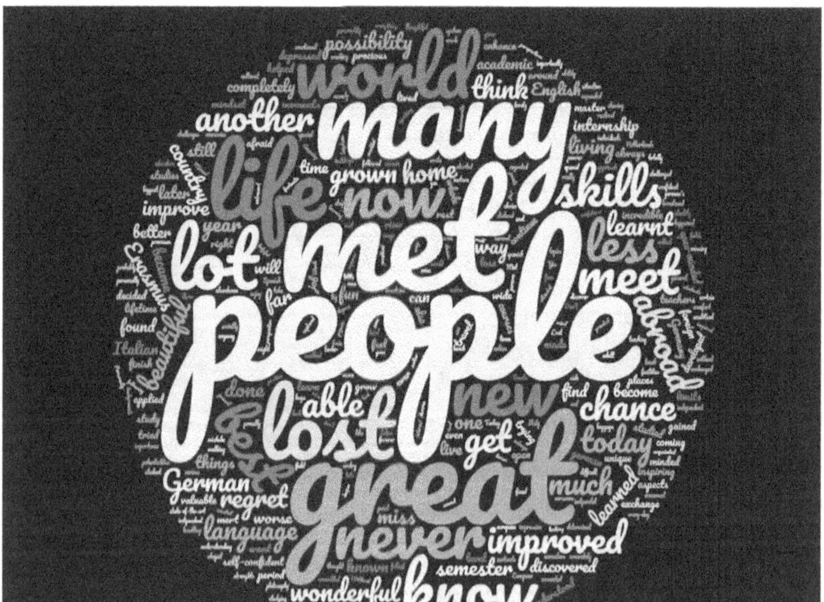

Figure 8.1 Top-of-mind mentions about what the Erasmus mobility experience meant to participants.
Source: the authors' own elaboration using Wordclouds.com.

Table 8.4 Top of mind perception of Erasmus mobility outcomes

Codes	*Percent*
Opportunities	21.9
Personality development	19.6
Important experience	19.6
Meet many people	14.6
Explore different cultures	6.4
Language skills improvement	5.0
Make international friends	4.6
Doesn't know	2.7
Other skills development	2.3
Carrier improvement	2.3
Negative experience	0.9

Source: Authors' own elaboration.

> "would have not learnt how to live in the same house with different cultures" (a female from Italy)

Results are also evidenced in the wordcloud by using Wordclouds.com with the expressions that were mentioned at least ten times.

The Erasmus exchange contributes to establishing international networks in part by meeting people from other countries (14.6%) and by making friendships (4.6%). Getting in touch with other cultures (6.4%) and improving language skills (5.0%) were also mentioned (Table 8.4). This means that participants valued less relationship building capability than their personal development.

Some participants recalled it as a negative experience (0.9%) or did not know why the Erasmus exchange should have been meaningful for them. It is surprising, although, that gaining knowledge, enriching one's University curricula, or – to put it in one word – studying, did not appear in any of the 219 answers.

Emotions communicate information (Irimiás & Michalkó, 2016). Participation in Erasmus mobility influences mental-health status by eliciting a wide range of emotions. Students were asked to rate on a 5-point Likert scale 16 Affectometer_2 items – 9 positive and 7 negative – presented as adjectives describing how often they felt in a certain way during Erasmus (Table 8.5).

All positive items scored significantly higher than any negative item, which means that being abroad with Erasmus is overall a positive experience. Among negative items, 'insecure' (2.32) and 'misunderstood' (2.23) scored higher but none of those reached an average of three.

Interestingly, mental-health states of being 'relaxed' and 'happy' showed some peculiarities among age groups. Younger participants (19–21 years) compared to the responses of older participants (22–28

Table 8.5 Adjectives that most described participants' mental-health state while on Erasmus

Affectometer_2 items	*Mean*	*Std. Dev.*
Interested in other people	4.45	0.692
Happy	4.25	0.688
Excited	4.23	0.706
Proud	4.04	0.861
Understood	3.90	0.610
Self-confident	3.85	0.730
Empathetic	3.67	1.024
Relaxed	3.63	0.848
Useful	3.61	0.790
Insecure	2.32	0.952
Misunderstood	2.23	0.833
Afraid	1.98	0.821
Homesick	1.89	0.839
Helpless	1.86	0.920
Depressed	1.69	0.757
Ashamed	1.65	0.753

Source: Authors' own elaboration, 2019.

years) felt significant more often to be 'relaxed' (4.11 vs. 3.47 and 3.65) (F = 10.34; p = 0.000), and 'happy' (4.59 vs. 4.21 and 4.02) (F = 8.48; p = 0.000). Respondents who spent a longer period on Erasmus (at least 9 months) reported significantly more often being 'depressed' (1.98 vs. 1.58) compared to those who spend only 3–5 months abroad (F = 4.79; p = 0.006). Intercultural sensitivity and humanistic management are associated with empathy. It is surprising that being 'empathetic' – perceiving emotions of others and behaving according to another person's needs – scored only 3.67. All the six positive mental-health statuses – from being 'interested in other people' to being 'self-confident' – are narcissistically concentrated on the self. Being 'useful' –another prosocial behavior – scored even lower (3.61) than being empathetic. To test which mental-health statuses influenced the perceived relevance of Erasmus, experience for future career regression analysis was used.

Overall, the participants affirmed that the Erasmus-mobility experience was relevant for their career (Mean 6.41; SD 1.06), which underlines the importance of such study-abroad programs. There is no significant correlation between the perceived usefulness of Erasmus study-abroad programs and the variables of age and gender. In order to see how emotions and mental-health issues influenced the perceived usefulness of the Erasmus study-abroad program, linear regression analysis was carried out. The results of regression analysis, (R^2 = 0.253; F = 24.23; p = 0.000) showed that from the 16-items, only three items – 'helpless' ($ß$ = –0.277; t = –4.569; p = 0.000), 'happy' ($ß$ = 0.241;

$t = 3.801$; $p = 0.000$), and 'self-confident' ($ß = 0.207$; $t = 3.209$; $p = 0.002$) – were statistically significant to influence the evaluation of the Erasmus experience as worthwhile for a career (Stepwise, Criteria: Probability-of-F-to-enter < = .050, Probability-of-F-to-remove > = .100). Being happy and self-confident was associated with higher usefulness. In contrast, being often helpless resulted in perceiving the study-abroad experience as useless (Table 8.6).

The ethnocentric stages of DMIS – denial, defense – were absent from the participants' self-evaluation answers. This result was expected because students of management and tourism on an international exchange program are usually willing to engage with people of diverse cultural backgrounds. In fact, most respondents affirmed to have intentionally sought to act with culturally different people (Mean 5.63, SD 1.336), and the confidence to establish contacts with fellow foreigners is also confirmed. The perceived security participants recalled saying that in a multicultural environment, their cultural values were not threatened (Mean 1.75, SD 1.224).

We found support for the contention that minimization of cultural differences was widespread among students: "I believe that when emotions are expressed those are universally recognizable" (Mean 5.38, SD 1.707). These results are in line with the findings of Reisinger and Crotts (2010), confirming that being in an intercultural environment intersects subjective cultures to homogeneity. Ethnorelative stages – acceptance, adaptation, and integration – reflected more how participants evaluated their own intercultural sensitivity (Table 8.7). Items measuring the

Table 8.6 Coefficients on mental-health status and the evaluation of Erasmus Experiences as worthwhile

Coefficients[a]

Model		*Unstandardized Coefficients*		*Standardized Coefficients*		
		B	*Std. Error*	*Beta*	*t*	*Sig.*
3	(Constant)	4.293	.505		8.494	.000
	Happy	.371	.097	.241	3.801	.000
	Helpless	–.319	.070	–.277	–4.569	.000
	Self-confident	.291	.091	.201	3.209	.002

Source: Authors' own elaboration, 2019.

Notes

a Dependent Variable: In terms of my career, I view my Erasmus experience extremely worthwhile.

ethnorelative stages of scored high, with participants recognizing the validity of multiple world views (Mean 5.60, SD 1.385) and believing that culture is a process in evolution (Mean 6.02, SD 1.088).

Students' managerial skills and competencies could be improved by Erasmus mobility. Although students' self-evaluation on their management skills might not be associated only with their Erasmus experience, the evaluation of such skills could have been reinforced by increased self-confidence. We used Olson & Kroeger's (2001) Intercultural Communication Scale, which was enriched with 6-items related to decision making that we deemed relevant to developing skills in humanistic management. Results did not evidence significant differences associated with the months spent on Erasmus or with nationality. Some items predict a narrow difference (e.g., female students are linguistically more competent), but this confirms the general assumption that females are usually better in communication (Table 8.8).

On the self-evaluation of intercultural communication skills, being linguistically and culturally competent scored significantly higher than any other skill. This result might be linked to the reason why participants decided to study abroad. In this set of items empathy – 'I have the ability to psychologically put myself into another person's shoes' scored higher (5.48) than the trust in co-workers (5.39). The low score of the 'being confident to have good leadership skills' (4.67) is an outcome linked to the young age of respondents who, as it is assumed, had limited opportunities to test their leadership skills. Students' low self-esteem is a relevant finding that universities should consider more. Although studies in psychology have evidenced that high self-esteem does not predict better academic or job performance (Baumeister et al. 2003), people with a healthy self-esteem manage social interactions better and are more persistent after failure.

5 Conclusions

Findings show that students consider themselves ethnorelative and consider different worldviews a value, something to respect and to endorse. On one hand, this is significant result in students' self-evaluation, on the other it provides a good humus to become culturally responsive managers and citizens. In an increasingly interconnected and highly demanding business environment self-awareness along with the respect of diversity predicts that a humanistic turn is possible. To improve humanistic management ethical principles to recognize the value of diversity is the way to a correct understanding of human values. Although minimizing cultural differences, assuming that one's set of norms and values are universal, is diffused among students they try to avoid stereotyping and prejudice. Our results confirm Anderson et al.'s (2006) findings that Erasmus study-abroad programs are perceived by participants as useful

Table 8.7 Descriptive statistics on students perceived intercultural sensitivity

Selected items/constructs	*Mean*	*SD*
I do not really notice cultural differences.	3.20	1.674
I have intentionally sought to make contacts with culturally diverse people	5.63	1.336
In a multicultural environment I feel like my cultural values are threatened	1.75	1.224
Some general behavioural rules are valid in every culture	5.11	1.499
I believe that when emotions are expressed those are universally recognizable. (eg. A smile is a smile wherever you go.)	5.38	1.707
I respect cultural differences	6.52	.774
I believe that behaviours vary across cultures	5.71	1.119
I believe that my worldview is one of many equally valid worldviews	5.60	1.385
I have added to my own cultural skills new skills from another culture	5.70	1.219
I believe that culture is a process, it can evolve	6.02	1.088
I feel positive about cultural differences	6.05	1.003
I feel to be a bridge between different cultures	5.11	1.434
I can consciously make a choice to act according to the cultural contexts	5.36	1.071

Source: Authors' own elaboration based on the Intercultural Sensitivity Scale (Olson & Kroeger, 2001), 2019.

to develop critical consciousness about issues of cultural diversity and social justice.

From students' self-evaluation on intercultural sensitivity and communication skills, it emerged that universities should invest more in intercultural training sessions. We think that it would be advantageous to involve Erasmus students in standard courses to stimulate interaction with home students and to provide immersive experiences for the benefit of home and international students. From the evaluation of the support of higher education institutions, it was evidenced that universities should provide training sessions on interculturality before departure and after returning from the international study program to stimulate reflection on the experience, and on the self-identity as well. It has been acknowledged that getting into contact with different cultures, *per se*, fails to increase intercultural sensitivity.

The limits of this study derive from the fact that no pre-survey was administered; thus it cannot be assumed that the students' self-perceptions about their intercultural communication and linguistic competencies were due to the Erasmus-mobility experience. These could have been pre-existing. For future research, we suggest developing a new

Table 8.8 Descriptive statistics on intercultural communication and management skills developed due to the Erasmus study-abroad programs

Items/constructs	*Mean*	*SD*
I am linguistically and culturally competent in at least one language and culture other than my own.	6.16	1.067
I try to learn about people from other cultures so that we can work and socialize together.	6.11	0.907
I appreciate how people from other cultures are different from me.	5.94	1.034
I have good problem-solving skills.	5.58	1.152
I feel trusted.	5.55	1.142
I have the ability to psychologically put myself into another person's shoes	5.48	1.198
I incorporate the attractive aspects of other cultures into my own way of doing things.	5.42	1.192
I have trust in people who work with me.	5.39	1.166
I am able to motivate people.	5.20	1.333
I have substantive competence in analyzing global issues.	5.15	1.165
I can act as a cultural mediator and serve as a bridge between people of different cultures.	5.15	1.350
I am confident that I have good leadership skills.	4.67	1.512
I have high self-esteem.	4.63	1.598

Source: Authors' own elaboration.

scale based on Bennett's model and on the Intercultural Sensitivity Scale for proper measurement of constructs of intercultural sensitivity.

6 Managerial Implications

This research indicates the capacity of Erasmus study abroad programs ranging from 3 to 9 months to have an impact on students' self-perceived intercultural sensitivity. Our results are consistent with previous findings by Anderson et al. (2006). Based on results revealed by the Affectometer_2, we recommend sending and hosting institutions to support Erasmus students by providing them with assistance to reduce the perceived status of being 'helpless' and 'useless' and to let students build upon the international experience for their management education. For this purpose, we suggest universities address the issue of intercultural sensitivity in their curriculum endeavoring to educate empathetic, ethical, and non-discriminative humanistic managers with intercultural competencies. As previous studies highlighted, the long-term effects of study-abroad programs are enduring if students are stimulated to reflect on their experiences (Nguyen, 2017). Thus, we propose setting up a humanistic management laboratory addressed to Erasmus and home students to develop intercultural sensitivity based on mindset. In the laboratory real-life cross-culturally situations – for example, diverse attitudes about being late, about storytelling, or about complaining – would be simulated in-class. MBA curricula in business and tourism management should be implemented by innovative human-centered teaching methods involving on-campus students as well (an interesting example is constituted by virtual short-term intercultural exchange lessons [Choi & Choi, 2020]). Students could practice and improve their communication skills and intercultural sensitivity by reflecting on different cultural attitudes and behaviors in problem-solving.

Faculty leaders should consider how the benefits of training interculturally sensitive humanistic managers can be integrated as part of the larger institutional goals. It is without any doubt that higher education and the business environment along with tourism are part of an international and global ecosystem. Intercultural competences developed by study-abroad programs and/or human-centered courses with a guided reflection on soft skills, ethics, and empathy can hopefully armor students to future challenges that we cannot foresee.

References

Anderson, P. H., Lawton, L., Rexeisen, R. J. & Hubbard, A. C. (2006). Short-term study abroad and intercultural sensitivity: A pilot study. *International Journal of Intercultural Relations*, *30*(4), 457–469. doi: 10.1016/j.ijintrel.2005.10.004.

Antal, A. B. & Friedman, V. J. (2008). Learning to negotiate reality: A strategy for teaching intercultural competencies. *Journal of Management Education*, *32*(3), 363–386.

Barron, P. & Dasli, M. (2010). Towards an understanding of integration amongst hospitality and tourism Students using Bennett's Developmental Model of Intercultural Sensitivity. *Journal of Hospitality, Leisure, Sport, and Tourism Education*, *9*(2), 77–88.

Baumeister, R., Campbell, J., Krueger, J. & Vohs, K. (2003). Does high self-esteem cause better performance, interpersonal success, happiness, or healthier lifestyles? *Psychological Science in the Public Interest*, *4*(1), 1–44.

Bennett, M. (1986). A developmental approach to training for intercultural sensitivity. *International Journal of Intercultural Relations*, *10*(2), 179–196.

Bennett, J. & Bennett, M. (2004). Developing intercultural sensitivity: An integrative approach to global and domestic diversity. In D. Landis, J. Bennett & M. Bennett (eds.), *Handbook of intercultural training* (pp. 147–165), Thousands Oaks, CA: Sage Publications.

Choi, S. H. & Choi, S. H. (2020). Virtual short-term intercultural exchange as an inclusive educational strategy: Lessons from the collaboration of two classes in

South Korea and China, *Journal of Teaching in Travel & Tourism*. Advance online publication. doi: 10.1080/15313220.2019.1707147.

De Mooji, M. (2019). *Consumer Behavior and Culture: Consequences for Global Marketing and Advertising* (3rd ed.). Thousand Oaks, CA: Sage.

Earley, P. C. & Ang, S. (2003). *Cultural intelligence: Individual interactions across cultures*. Stanford, CA: Stanford University Press.

Farmaki, A. (2018). Tourism and hospitality internships: A prologue to career intentions? *Journal of Hospitality, Leisure, Sport and Tourism Education, 23*, 50–58.

Giudici, E., Varriale, L., Floris, M. & Dessì, S. (2011). Teaching business students to be passionate about ethical sustainable development. In Charles Wankel, Agata Stachowicz-Stanusch (eds.), *Effectively integrating ethical dimensions into business education* (pp. 135–158). Charlotte, North Carolina: IAP-Information Age Publishing Inc.

Erasmus Higher Education Impact Study (2019). https://publications.europa.eu/en/publication-detail/-/publication/94d97f5c-7ae2–11e9-9f05-01aa75ed71a1/language-en.

Hammer, M. R. (2011). Additional cross-cultural validity testing of the Intercultural Development Inventory. *International Journal of Intercultural Relations*, *35*(4), 474–484.

Hammer, M. R., Bennett, M. J. & Wiseman, R. (2003). Measuring intercultural sensitivity: The intercultural development inventory. *International Journal of Intercultural Relations*, *27*(3), 421–443.

Hogg, M. & Terry, D. (2000). Social identity and self-categorization processes in organizational contexts. *The Academy of Management Review*, *25*(1), 121–140.

Irimiás, A. & Franch, M. (2019). Developing intercultural sensitivity as an emotional ability. In Koc, E. (eds) *Emotional intelligence in tourism and hospitality* (pp. 95–107), Wallingford: CABI.

Irimiás, A. & Michalkó, G. (2016). Hosting while being hosted: A perspective of Hungarian migrant hospitality workers in London, UK. *Travel and Hospitality Research*, *16*(2), 172–183.

Jackson, T. & Haines, R. (2007). Cross-cultural management in South African NGOs. *South African Review of Sociology*, *38*(1), 85–98.

Jackson, J. (2015). Becoming interculturally competent: Theory to practice in international education. *International Journal of Intercultural Relations*, *48*, 91–107.

Kenesei, Z. & Stier, Z. (2016). Managing communication and cultural barriers in intercultural service encounters. Strategies from the both sides of the counter. *Journal of Vacation Marketing*, *23*(4), 307–321.

Kirillova, K., Lehto, X. & Cai, L. (2015). Volunteer tourism and intercultural sensitivity: the role of interaction with host communities. *Journal of Travel and Tourism Marketing*, *32*(4), 382–400.

Lugosi, P. & Jameson, S. (2017). Challenges in hospitality management education: Perspectives from the United Kingdom, *Journal of Hospitality and Tourism Management*, *31*, 163–172.

Olson, L. C. & Kroeger, K. R. (2001). Global competency and intercultural sensitivity. *Journal of Studies in International Education*, *5*(2), 116–137.

Ozcelik, H. & Zoltay Paprika, Z. (2010). Developing emotional awareness in cross-cultural communication: A video conferencing approach. *Journal of Management Education*, *34*(5), 671–699.

National Health Scotland (2006). Measuring positive mental health: Developing a new scale. http://www.healthscotland.scot/publications/measuring-positive-mental-health-developing-a-new-scale (last accessed on 27 July 2019).

Nguyen, A. (2017). Intercultural competence in short-term study abroad. *Frontiers: the Interdisciplinary Journal of Study Abroad*, *29*(2), 109–127.

Patterson, P. K. (2006). *Effect of study abroad on intercultural sensitivity* (Doctoral dissertation, University of Missouri-Columbia).

Peng, A. C., Van Dyne, L. & Oh, K. (2015). The Influence of motivational cultural intelligence on cultural effectiveness based on study abroad: The moderating role of participant's cultural identity. *Journal of Management Education*, *39*(5), 572–596.

Pirson, M. (2019). Editorial: Managing towards a world that works for all. *Humanistic Management Journal*, *4*, 1–4.

Pirson, M. (2018). Reclaiming our humanity- a cornerstone for better management. *Humanistic Management Journal*, 2, 13–107.

Pirson, M. (2017). *Humanistic management. Protecting dignity and promoting well-being*. Cambridge: Cambridge University Press.

Reisinger, Y. & Turner, L. (2011). *Cross-cultural behaviour in tourism: Concepts and analysis*. New York: Routledge.

Reisinger, Y. & Crotts, J. (2010). Applying Hofstede's National Culture Measures in tourism research: Illuminating issues of divergence and convergence. *Journal of Travel Research*, *49*(2), 153–164.

Reisinger, Y. & Turner, L. (1997). Cross-cultural differences in tourism: Indonesian tourists to Australia. *Tourism Management*, *18*(3), 139–147.

Reisinger, Y. & Turner, L. (1998). Cultural differences between Mandarin-speaking tourists and Australian hosts and their impact on cross-cultural tourist host interaction. *Journal of Business Research*, *42*, 175–187.

Rexeisen, R. J., Anderson, P. H., Lawton, L. & Hubbard, A. C. (2008). Study abroad and intercultural development: A longitudinal study. *Frontiers: The Interdisciplinary Journal of Study Abroad*, *17*(1), 1–20.

Shaftel, J., Shaftel, T. & Ahluwalia, R. (2007). International educational experience and intercultural competence. *Journal of Business & Economics*, *6*(1), 25–34.

Sizoo, S., Iskat, W., Plank, R. & Serric, H. (2003). Cross-cultural service encounters in the hospitality industry and the effect of intercultural sensitivity on employee performance. *International Journal of Hospitality and Tourism Administration*, *4*(2), 61–77.

Triandis, H. (1994). *Culture and Social Behaviour*. New York: McGraw-Hill.

Triandis, H. (2006). Cultural intelligence in organizations. *Group and Organization Management*, *31*(1), 20–26.

Yurur, S., Koc, E., Taskin, C. & Boz, H. (2018). Factors influencing intercultural sensitivity of hospitality employees. *International Journal of Hospitality & Tourism Administration* doi: 10.1080/15256480.2018.1547236.

Westrick, I. (2004). The influence of service learning on intercultural sensitivity. A quantitative study. *Journal of Research in International Education*, *3*(3), 277–299.

UNWTO International Tourism Highlights (2019). https://www.e-unwto.org/doi/pdf/10.18111/9789284421152.

9 A New Entrepreneurial Training Model

The "Win-Win UNESCO Experience"

Monica Basile

1 Introduction

Over the last few years, the tourism industry has seen a drastic evolution: it is clear just taking into account the social and economic changes caused by the introduction of new technologies (in particular the OTAs or Online Travel Agencies), by the boom of low-cost flights (Basile, 2009a), and by the wider availability of people's leisure time and financial resources. These facilitations have undoubtedly made several tourist destinations more accessible, while creating, at the same time, a phenomenon of marginalization of others, since most of the tourist flows are focused only on some mainstream choices. The competition amongst the most popular destinations has increased significantly compared to the less-visited ones (Crouch, 2011), in which tourist products are less diversified and therefore weaken their competitiveness on the global market.

Therefore, along with an entire sector of distressed touristic destinations, there is also a more transversal sector, in which seasonality plays a vital role: both in well-known or less popular sites, touristic flows can reduce drastically or stop entirely based only on the seasons, having negative consequences to the entire entrepreneurial, economic, and social context of the area.

These events can be interpreted as being less negative when the rise of new types of tourists is taken into consideration: we are referring to pensioners, cultured people with children to whom they want to impart a conscious approach to cultural and ecological values (Vidal González, 2008). These cultural tourists have "the intention to gather new information and experiences to satisfy their cultural needs" (Richards 1996, p. 24), they represent 40% of the total international arrivals (UNWTO, 2015a), and they are constantly growing (UNWTO, web sites). Recently a new type of "creative tourists" is emerging, consisting of those who are more interested in developing "their creative potential through active participation in courses and learning experiences which are characteristic of the holiday destination where they are undertaken" (Richards & Raymond, 2000, p. 30). Creative tourists are mainly focused on discovering the "intangible heritage" of a place (characterized by the residents' habits, custom, and lifestyle; creativity and media; identity and narratives), as well as discovering the "tangible heritage" (characterized by museums, monuments, artistic and natural beauty of the area) (Richards & Wilson 2007, p. 18).

In a nutshell, culturally connoted tourism craves to live authentic and significant experiences in unique places – which are often identified as UNESCO destinations (natural, cultural, landscape, and creative cities) – and in which getting in touch with local inhabitants is considered to be an added value (Richards & Wilson 2007).

In order to give effective answers to this new tourist demand, it is necessary that the offer is made by rethinking its role within the tourism system, along with its own product, which are still too often grounded to models based on the development of mass tourism (Vanhove, 1997), and nowadays even inappropriate and just a source of problems, such as in the case of over-tourism. The demand of these new tourists represents an opportunity for the introduction of new supply models that increase the overall quality and convey it through the uniqueness of the product-experience of the tangible and intangible heritage of the destination (Valentino, 2001).

In order to introduce these innovation processes at any given destination, international institutions UNESCO, ICCROM, ICOMOS and IUCN (2013), UNWTO (2015b), UNDP (2010), UN 2030 Agenda (sitography) underline the importance of fostering the opportunities of "capacity building" in order to build stronger public–private stakeholders.

Capacity-building introduces an inclusive approach, based on the development of relationships between local stakeholders and on people-centered change dynamics, to achieve together a higher common goal of preserving and improving the management of the UNESCO World Heritage Site (WHS)[1] by creating positive effects for the individual, for local populations, and for society. This sustainable development approach adopted by international organizations, through the tool of capacity building, is very close and linked to the principles of Humanistic Management (HM).

However, the literature does not give any specific capacity-building model for local multi-sectorial entrepreneurs in WHS's destinations that want to develop a long-lasting partnership to enhance the tangible and intangible elements of heritage, which lead to the creation of a sustainable tourism product; even the academy affirms that there is still little research concerning capacity-building related to the tourism sector (Moscardo, 2008). At the same time, HM also affirms the importance of collaboration among businesses in order to achieve common goals for the common good of society and indicates that there is a need to further deepen the theory with practical new studies (Melé, 2003).

This chapter aims precisely to fill this gap, presenting a new type of training model, called the "Win-Win UNESCO Experience", based on the inter-company and cross-sectoral fusion between public-private stakeholders within destinations who, during capacity-building, work together, spontaneously activate innovative processes, which, by the end of the training session, allow for the creation of a new touristic product-experience or cultural itinerary, which is self-sustaining even up to 10 years after its introduction. This model has been designed, developed, and applied by the author herself in World Heritage Sites (WHS) and has generated results that are still withstanding. In particular, it shows that the "bottom-up" approach is the most suitable one for an entrepreneurial audience since it allows for training needs to emerge, which can then be linked up during the creation of specific products targeted to the "progressive" target of the experiential tourism.

2 Theoretical Background

The theoretical background includes several theories based on capacity-building, community capacity-building, and theories linked to HM, which activate spontaneous processes of entrepreneurial, corporate, and social innovation.

2.1 *Capacity Building*

Capacity buildings (CB) is indicated both by international organizations as well as by the academy as an important tool of sustainable development to support the well-being of populations, society and to enhance the local tangible and intangible heritage. CB means building the capacities of an organization to meet the demands of change. CB is an ongoing process of improvement and strengthening of the already-existing skills.

On the other hand, CB was also adopted by UNESCO, ICCROM, ICOMOS, & IUCN (2013 p. 50) with the document "World Heritage Strategy for Capacity-Building", in order to introduce effective management for World Heritage Sites. Capacity-building – whether by practitioners, institutions, or communities and networks – is seen as a form of people-centered change that entails working with groups of individuals to make improvements in approaches to managing cultural heritage.

Capacity Building is also a pillar of the UN 2030 Agenda for Sustainable Development. It is mentioned among the Goal 17, which aims at revitalizing the public-private partnership to engage multiple stakeholders. In particular, it is quoted in target 17.9, which aims to "enhance international support for implementing effective and targeted capacity-building in developing countries to support national plans to implement all the sustainable development goals". Capacity building is also seen as an Agenda 2030 "Partnership accelerator" in order to "build the capacity of relevant stakeholders to develop and implement partnerships for the SDGs, and to support organizations to develop their policy and strategy, systems and processes, legal agreements and culture to support collaboration" (sitography).

International institutions confirm the importance of achieving SDGs with Goal 17, by fostering inclusive Public-Private Partnerships (PPPs) (UNWTO 2015b), without, however, illustrating how these models could be effectively implemented in detail.

Capacity-building and community capacity-building are closely associated terms.

The community capacity-building (CCB)-based approach has become increasingly important to help communities participate in development initiatives, including tourism. The lack of community capacity and limited understanding of tourism impacts have been recognized as barriers to effective tourism development. Unfortunately, the concept of CCB applied to tourism development has not been deeply analyzed yet, while it has been extensively used for other sectors, such as agriculture, education (Smyth, 2009), and health (Maclellan-Wright et al., 2007). No specific definition has been given for CCB focused on stakeholders or communities with respect to tourism (Aref & Redzun, 2008). There is very little literature that discusses the practical application of approaches that have been successfully used to measure barriers to CCB for tourism development (Moscardo, 2008).

The chapter explores the structure of CB that develops the methodologies of "co-creation, co-design, or co-production of knowledge, and/or systems and/or outcomes. The emphasis here is on bottom-up discourse formation, as opposed to traditional authoritarian, imposed definitions and parameters. Co-production aims for collaborative ways to create knowledge and address problems occurring in specific contexts (Fontan et al. 2013 in Tiwari, Lommerse, & Smith, 2014, p. 261).

2.2 *Humanistic-Centered Approaches*

The community-based tourism approach (CBT) is strongly associated with Humanistic Management based on public good and community well-being. Melé (2016, p. 48) states that "The stakeholder's approach (Freedman, 2004) sees the purpose of the firm in satisfying the interests of the stakeholders of the firm (shareholders, employees, customers, suppliers, and local communities, and often others), or creating value for them. This vision is closer to humanistic vision, although the relationship between the firms and its stakeholders are exclusively based on interests, and so it seems an individualistic view of society".

Scheyvens (1999, p. 11) identifies a number of critical success factors (CSFs) for community-based tourism that are common worldwide which are highly correlated with HM and which "CBT thus very clearly identifies with distributive and social justice, ethical relationship and equity, from its rootedness in the local/community. In CBS, Industry is micro-level, practical, emphasis on local enterprises and local economic development, social well-being, cultural survival, poverty alleviation, capacity building.

3 The Model of "Win-Win UNESCO Experience"

The "Win-Win UNESCO Experience" model emerges from an integrated approach of the theories which have been previously illustrated. The model uses the Community Based approach, in which entrepreneurs are residents who have lived in WHS for generations, who play an active role in capacity-building planning. It also applies HM's principles based on respect for the individuals as well as the guarantee that dignity is protected during training. Capacity-building develops the participants' skills and empowerment, which are set up according to the methodology of the social and sociological research (Cipolla, 1996) that employs both quantitative tools (questionnaires, even anonymous ones, in order to facilitate the honest opinion of the participants) and qualitative tools (focus group, open discussion).

The framework within which the "Win-Win UNESCO Experience" model fits is WHS destination, in which local multi-sector entrepreneurs (food and wine, tourism, farmers, handicraftsman, creative industries, etc.) rarely dialogue in a systematic and organized way either among themselves or with local cultural institutions. The aim of the "Win-Win UNESCO Experience" model is to introduce CB in order to develop relationships based on trust between the stakeholders of the public sector (i.e., Museum, Institutions, World Heritage Managers, etc.) and the private sector (Entrepreneurs from different economic branches) of a World Heritage Site, with the purpose of triggering spontaneous personal, social, business innovation processes that generate the creation of experiential tourist products (Richards & Wilson 2007) and cultural itineraries that self-support themselves over time, which are referred to as the "Win-Win UNESCO Experience". During CB, the private-public stakeholders get to know one another and establish a permanent partnership, sharing common objectives, for example, relaunching the tourist destination by creating a cultural product-experience that attracts cultural and experiential tourists as an alternative to mass tourism (Scheyvens, 1999).

It is an innovative long-term local-development model, based on the active role played by the local entrepreneurs of the destination by promoting their cultural Heritage. The challenge is represented by designing a capacity-building strategy that is considered interesting and effective for both public and private stakeholders. In fact, entrepreneurs are characterized, in most cases, by a concrete approach based on short-term quantitative results (Scheyvens, Banks, and Hughes 2016); on the other hand, cultural institutions tend to protect and preserve heritage, rather than promote it with entrepreneurs.

The model guarantees that at the end of the CB process, the public-private working group designs a product-experience based on the enhancement of the UNESCO heritage as well as on the revitalization of the intangible culture, linked to the local entrepreneurial know-how.

The author provided a direct contribution to the capacity-building model: she set up the model that leads to the creation of the sustainable tourist products-experience based on UNESCO heritage; she designed it based on academic and empirical knowledge learned in twenty years' working experience developing tourist destinations with local entrepreneurs; she worked with the client/institution to design and develop the specific project; she developed and managed all the "people", phases and contents concerning the capacity building in order to reach the objectives, she conceived and managed all the training sessions with entrepreneurs, she supervised the creation of the tourist product; she facilitated the introduction of innovative processes among enterprises and cultural institutions.

This scheme below (Figure 9.1), designed by the author for this chapter, illustrates the process and phases of "Win-Win UNESCO Experience" capacity building:

The process of "Win-Win UNESCO Experience" capacity building including the phases and contents for each phase.

a *Introduction of policies to support entrepreneurial training.*

Through the "top-down" approach, the financing Institutions (i.e., the local Government, Destination Management Organization -DMO's-, Chambers of Commerce, Employers Trade Organization, WHs Manager) introduce innovative tourism destination policies in order to add value to cultural heritage and World Heritage Site, by creating a "networked" touristic product and by financing the capacity buildings between the local public-private stakeholders. Tourism policy supports the creation of a network developed among local businesses and between enterprises and cultural institutions, in order to enhance the WHS tangible and intangible culture through entrepreneur activities. Applying the "top-down" approach, the public stakeholders implement innovation processes, by encouraging small and medium-sized enterprises to cooperate together (Mariani, Baggio, Buhalis, and Longhi 2014). Several studies state that innovation processes cannot occur in isolation, but they are more likely to take place when networks of individuals or companies work together (Sawyer, 2009 in: Mariani et al. 2014).

b *Establishing results to be reached on behalf of the stakeholders and of the destination.*

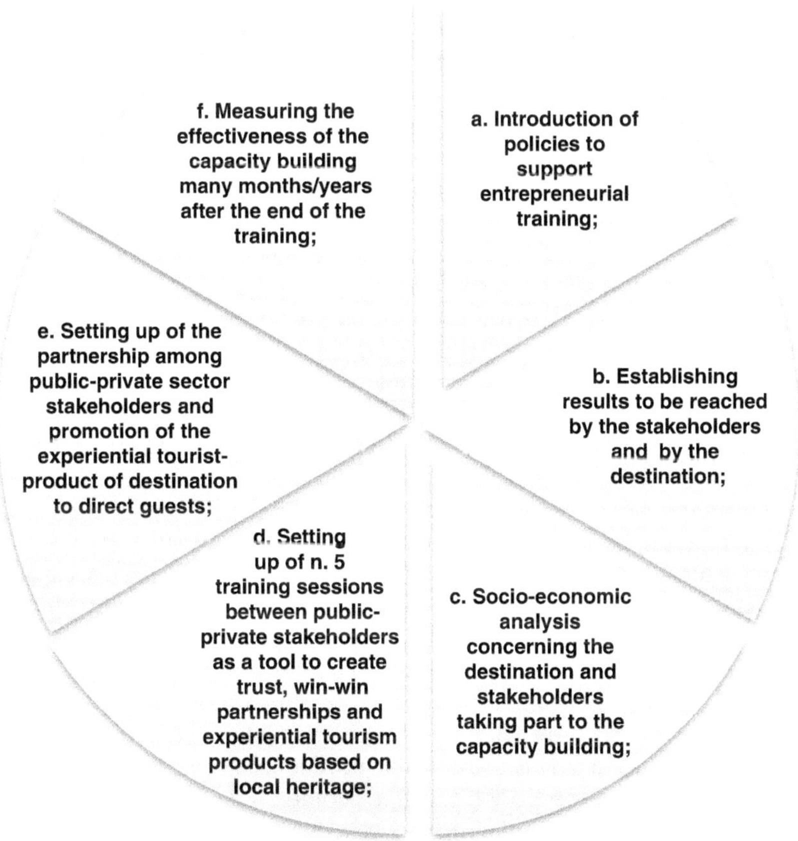

Figure 9.1 "Win-Win UNESCO Experience".
Source: Author's own elaboration.

The institution decides to introduce an external support, to activate the capacity-building process. In this phase, the trainer receives the objectives from the financing Institution which have to be achieved through the training project. The lecturer also provides the Institution with the following consulting services: advice on strategies to be introduced to inform and motivate local entrepreneurs to take part in capacity building; technical assistance for strategy and communication in order to effectively manage and strengthen the relationship with entrepreneurs, the design of specific training for the Institution's employees in order to stimulate the participation of the most innovative local entrepreneurs; advice on strategies to be introduced at the destination, in order to build consensus on the capacity-building project among public-private stakeholders and opinion leaders; design the content and phases of the training course in order to ensure the achievement of the results agreed upon with the Institution. Milen, (2001 p. 1 in Loza, 2004 p. 302) among the effective capacity-building practicing genuine partnerships, states that "this is a consultative and collaborative process and requires strong communication within partners and alliances to ensure that all stakeholders are fully aware of the capacity-building program, the needs identified ant the expected outcomes".

c *Socio-economic analysis concerning the destination and stakeholders taking part in the capacity building.*

The preparatory phase and the analysis of the specific local, social, and economic contexts, which play an extremely critical role in the success of the model. The trainer takes into account the previous tourism analysis carried out on the destination; the activity analysis of each entrepreneur who takes part as to establish their degree of business development through indicators of "web reputation" and through their website analysis; the potential partnership that could potentially be developed among stakeholders.

Milen, (2001 p. 1 in Loza, 2004 p. 302) among the effective capacity building understanding the context specificity of capacity and its development states to "Clearly define the question: capacity of what? Ensuring it is relevant to the mission statement of the organization". In this model, the competence to be acquired through the capacity building is the knowledge among private-public stakeholders to work together and to create a sustainable product based on the enhancement of the local intangible and tangible heritage.

d *Setting up of n.5 training sessions between public-private stakeholders.*

The seminars represent an essential tool of the model through which relationships of trust and Win-Win partnerships are established among the multi-sectorial stakeholders and experiential tourism products are designed based on local heritage.

The methodology is based on the participatory model "bottom-up" approach, which is mainly used to develop touristic-experiential products created by stakeholders during the training seminars, or capacity buildings. Entrepreneurs receive the adequate know-how and the useful tools from the lecturer/trainer so they can elaborate on the solutions to the problems found, starting from the bottom.

The pedagogical methodology adopted during the capacity building sessions, applied the Humanistic Management approach, assuring the well-being of the participants and the development of the general welfare of resident entrepreneurs (Pirson 2017). Furthermore, in an entrepreneurial lifelong learning perspective, this typology of training guarantees fair opportunities for personal development throughout their lifetime and provides to the participants the ability to acquire cognitive tools (Pirson 2017). They are therefore included in the "peer-to-peer" network of resident entrepreneurs, spreading, thereby, their personal and professional opportunities of expression (Pirson, 2017) protecting their individual dignity, and encouraging personal development.

The lecture is not structured in a classical manner with the professor teaching the entrepreneurs, but the trainer in this case plays a plurality of roles including that of coach, motivator, facilitator, mediator, and finally lecturer in order to clearly outline the new market trends.

This training model is structured in five main capacity-building sessions, which include: the negotiation with entrepreneurs concerning the business goals to be achieved; the knowledge of the World Heritage Site and of the services provided by the public-private stakeholders participating in the project; the development of "Win-win partnerships"; the creation of the experiential tourism product with the identification of a high standard of quality services shared among the partners; the promotion of the product through the websites of the participants.

e *Setting up of the partnership among public-private sector stakeholders and promotion of the experiential tourist-product of destination to direct guests.*

During this phase entrepreneurs work together, without the intervention of the institution or trainer, in order to concretely set up the partnership designed during the capacity building. Business owners also develop innovative processes within their firms by promoting and selling directly to their guests the product-experience created in the classroom. If the destination DMO (UNWTO, 2019) of the destination is involved in the project, it could contribute significantly to the success of the capacity building by promoting the tourist-experience product through web marketing actions in order to attract potential cultural-experiential customers.

f *Measuring the effectiveness of the capacity building many months/ years after the end of the training.*

In this phase, a new capacity-building takes place after a few months or years after the end of the training. The instructor goes back to the destination in order to evaluate the effectiveness of the partnerships developed autonomously by entrepreneurs. During the sixth capacity building, the trainer applies quantitative and qualitative methods in order to analyze the effectiveness of the project. Positive and negative constraints to the setting up of the partnership are discussed with entrepreneurs. The trainer and the institution further develop the project by re-designing new objectives to be achieved and improvements to be introduced to the model. The project is based upon entrepreneurs' motivation and passions, which guarantee its self-perpetuation over the years.

4 Case Studies of a Cultural Touristic Product "On the Etruscan Routes": Application of the "Win-Win UNESCO Experience" Model

Here is a case study in which the capacity-building model is applied to the creation of a tourist product in a WHS "St. Francis and other Franciscan places" (Italy, WHS 2000). In this case study attention will be focused on the introduction of the "Win-Win UNESCO Experience" training method applied to a cultural WHS. More precisely, it concerns the province of Perugia, well-known for the presence of "San Francis and other Franciscan Sites (WHS 2000) as well as the Places of Power (568–774 A.D.) (WHS 2011).

The project started from a successful capacity building introduced by the Association of Young Hoteliers of the Hotel Association Federalberghi Perugia trained by the author during the years 2014, 2015, and 2016. During the training, the entrepreneurs worked on destination management strategies by applying the "Win-Win UNESCO Experience" and creating tourist packages based on the enhancement of the intangible local heritage. During the following year, the author trained, with the same model, all the Italian Chamber of Commerce Tourism Managers; the capacity building took place in Rome and was promoted by the Italian Chambers of Commerce and ISNART. On that occasion, several dozen of Managers had the opportunity to learn the model and to practice it by creating heritage tourism products. Among the other destinations, the delegates of the Perugia Chamber of Commerce, subsequently through ISNART, decided to activate a long-lasting local development policy through the introduction of the capacity building (Phase a.).

The Chamber provided a socio-economic framework to the trainer, communicating that the destination is strongly characterized by a plurality of different handicrafts and entrepreneurial economic sectors (i.e., cashmere laboratories, chocolate stores, majolica potteries De Ruta) that do not interact with each other in a constant way. Through this capacity building the Chamber of Commerce aimed at creating long-term relationships among the multi-sectorial business activities and the cultural institutions (Phase b.).

The firms that could be eligible to participate in the capacity building had been identified by the Chamber of Commerce, taking them from the multi-sectorial database of businesses that were previously registered. The trainer gave advice on communication strategies and trained internal staff in order to have several highly motivated entrepreneurs to participate in capacity building.

During Phase c., priority to the capacity building, the trainer analyzed the local socio-economic context, as well as the tourism phenomena in the destination. She also deepened the investigations, studying in detail the activities of the n. 25 firms that were to take part in the capacity building, through the analysis of their websites and their web reputation. She formulated some hypotheses of win-win partnerships that could potentially be developed during the capacity building. The trainer designed through a "reverse-engineering process" a capacity-building content hypothesis, starting from the final objectives that the trainer had to achieve by the end of the project.

The beginning of Phase d., designing the capacity building, was based on the analysis of the participants' "training needs". The trainer designed an anonymous questionnaire in order to obtain, at the beginning of the first training session, a clear picture of the state of the art concerning the partnerships which had already been up by entrepreneurs. The questionnaire also investigated the business problems and the solutions the entrepreneurs wanted to achieve at the end of the training, as well as the resources that they were ready to invest in order to develop the multi-sector partnerships. The questionnaire was composed of n. 5 sections. The first one investigated the already existing presence of the multi-sectorial partnerships among the participants (with closed questions "yes/no"), with two open questions concerning the possibility of writing successfully and the failure of previously developed partnerships. The second section investigated the main business problems that motivated entrepreneurs to participate in capacity building. The entrepreneur was asked to express their opinion using a 1-9 scale vote, with number 1 indicating "I don't agree at all with the statement", and 9 "I totally agree with the statement".

4.1 Results of the Project

The capacity-building sessions (Phase d.), for a total duration of 16 hours, were based upon the questionnaire entrepreneurial needs, and they reached the following results.

The entrepreneurs spontaneously introduced a process of social, *organizational and product* innovation: afterward the hoteliers – that despite knowing each other for decades had never collaborated together before this training – decided to follow up the work they had already begun with amazing enthusiasm by mapping the available services, inciting their peers (who did not provide the proper information) to actively participate.

A subgroup represented the "Win-Win UNESCO Experience" product, created during the training, by illustrating an advertising brochure, with pictures and texts explaining the entrepreneurial services provided, in order to allow all the stakeholders of the network to promote the low season of Christmas 2018. The brochure includes twelve experiential activities with services and reduced prices, created by the participants of the training. Furthermore, during the last capacity building, the entrepreneurs spontaneously designed a cultural itinerary "On the Etruscan routes", which includes the tangible, intangible and entrepreneurial heritage of the participants and the destination; both were tested during the seminar by simulating the joint promotion methodology based on sharing mutual customers to the "colleagues" within the network, by guarantying a high-quality-standard service.

The effective training results reached by the capacity building highlights the importance of mediation between the stakeholders of different projects; it's quite rare, actually, that entrepreneurs start up a network on their own as a means to reach a common objective in a very short time, which has been identified, negotiated and shared within the training session (Basile, 2019). Even the anonymous final evaluations (Phase f.) of the entrepreneurial training course clearly show their great satisfaction at the end of the seminar

The anonymous questionnaire with evaluation scale 1 (minimum)–9 (maximum) were submitted to 21 entrepreneurs participating, which represented the 12 enterprises included in the promotional brochure. Results: average score: overview evaluation of the capacity buildings (8.3), effectiveness of the training course (8.2), organization (8.1), lecturer expertise and clarity of lecture exposition (8.6). Qualitative analysis concerning the sentiment of participants, processed through the semantic analysis highlighted the following results: participants could choose 3 out of 20 possible adjectives (between extremely positive and negative) to indicate their impressions concerning the capacity buildings. 72.6% of them chose: "productive, enlightening, motivating, stimulating, interesting"; 27.4% chose "new, reflective, useful, beautiful, rich, and logical". Source: Chamber of Commerce of Perugia

The Chamber of Commerce organized a public event to disseminate the results, and the trainer wrote an article describing the capacity-building experience, which were published on the Chambers' web site.

Among the long-term results of the project we include the following output. Considering the positive final results of the 2018 capacity building in Perugia, the Chamber of Commerce has decided to continue investing further resources on the "Win-Win UNESCO Experience" capacity-building project even for 2019. It planned three seminars to further develop the project of Perugia's network, which was created in 2018; moreover, it has decided to introduce the capacity-building project to two new destinations: Lake Trasimeno and Spoleto (WHS 2000), dedicating ten capacity buildings for the two new projects which lead to the design of the experiential offering http://www.unicaumbria.it/landing/itinerari-artigianali-e-culturali-spoleto/. http://www.unicaumbria.it/landing/itinerari-artigianali-e-culturali-trasimeno/

Concerning the Perugia's network, during the first capacity building, that took place in June 2019, after six months from the end of the training activity, the trainer invited the participating stakeholders (entrepreneurs and members of the local Chamber of Commerce) to carry out an "efficacy analysis" based on the results reached by the project during its first six months of life. The analysis investigated the real partnerships that entrepreneurs developed during six months (Phase e.), and it showed the true effectiveness of the network created in 2018. It was carried out both with social research qualitative and quantitative anonymous tools (Phase f.), which openly outlined the strengths and potential improvements.

Concerning Phase e., -the development of the partnership among stakeholders-, the qualitative analyses clearly highlighted the four innovation processes – social, relational, organizational, productive - that the participants introduced independently at the end of capacity-building training, in order to promote all the experiential activities of their peers during the low season. One worth mentioning is the organizational innovations concerning the new marketing strategies initiated to promote the "Win-Win UNESCO Experience" product to guests. Hoteliers declared that they invested in developing win-win partnerships with other stakeholders also through the promotion of the experiential brochure to their guests. They promoted the brochure through their own websites, within a specific webpage, at hotel receptions, through a welcome letter placed in the hotel rooms and even by e-mail. Furthermore, the producers have also promoted the network's activities through their websites.

Here are some good examples that emerged from the qualitative analyses concerning win-win partnership based on relational and product innovation. In one case, an hotelier designed two promotional corners set up within the lobby of the hotel, one dedicated to jewelry, of the handicraftsman specialized in Etruscan gold, - which could even be purchased – and another dedicated to the products of a craft-beer entrepreneur. Some hoteliers also included the network's products colleagues (oil, wine, beer) in their restaurant menus and spontaneously, some of their customers who tasted and appreciated them, then went to visit the winery and brewery producers. Other hotel guests, decided to create their own personalized chocolate at the producer's store and discover the UNESCO heritage through a visit to the churches.

Concerning Phase f., measuring the effectiveness of the capacity building after six months from the end of the training, from the quantitative analysis (in the form of anonymous questionnaires), which were handed out to the participating entrepreneurs, it emerged that after the 2018 capacity buildings brought to the following results: 100% of them started to develop new alliances with their colleagues, 85,7% sent their own customers to other entrepreneurs within the network, 57,1% increased their overall income by between 5% and 10% during the low season of the Christmas period (which was the goal that had to be reached by the end of the training sessions). Furthermore, participants stated that, during the sessions of capacity building, they acquired skills and expertise that would help them: "work together with new entrepreneurs who I did not know before, and reach a mutual goal" (average mark 8,9); "understand how other entrepreneurs can represent a new sales channel" (average mark 8,7); "start immediately to develop synergies with other entrepreneurs" (average mark 8,4); "play an active role within the creation of the product" (average mark 8,3); "relaunch my business activity due to the innovative marketing and sales strategies immediately applicable" (average mark 8,3); "develop my activity in collaboration with other entrepreneurs" (average mark 8,2); "understand the new needs of customers and create new services" (average mark 7,9); "increase the total income of the business" (average mark 7,7).

Concerning the open question "please state the most useful aspects that the training provided after six months", re-elaborated with the semantic analysis, the participants expressed the following motivations: 57.2% stated "the close cooperation among entrepreneurs, the exchange of ideas among businessman" and the 42.9% pointed out "the opportunity to get to know new entrepreneurs coming from different economic sectors"; on the other hand, to the open question "please, state the worthless aspects of the training", no participants expressed any negative comments.

To the open question (to which only some participants answered) asking "what did you like the most about this experience of working within a business network?" the answers were the following: "I was impressed by the real potential and the willingness to build up a group and to cultivate it in a way that it will remain active even in the future", "The opportunity to know new activities and finally the realization of a real network among businesses", "To get to know the views of every single entrepreneur who is part of the network", "The ability to discuss with entrepreneurs who want to reach mutual goals: the increase of the tourists flows and the high quality of the services provided", "I have had the opportunity to share and to compare opinions about topics of common interest". To the question "what did you like the least about the experience of working in a group?" there were just two answers: "The difficulties linked to the introduction of the changes within the enterprise", "Scarce participation of the hoteliers".

The goal set down by the network of private entrepreneurs (the participants of the five capacity-building sessions, financed by the Chamber of Commerce of Perugia and executed by the author through ISNART (Italian Research Institution for tourism by The Italian Chambers of Commerce), was the creation of a new product for the destination in order to relaunch the Christmas season of 2018. Surprisingly, the results went even further and, in addition to the achievement of the original goal, the trainees spontaneously generated (without any encouragement from the trainer) a cultural itinerary named "On the Etruscan routes" (Basile, 2019).

During the second year of the project, three new capacity building took place, a new business owner shared the common good of the project, and joined in. Capacity building led to the implementation of the cultural-experiential offering that was promoted through the Chamber of Commerce Unica Umbria web site. http://www.unicaumbria.it/landing/itinerari-artigianali-e-culturali-perugia/.

5 Conclusions

The main motivation of this article is to identify and to apply the HM approach as an important tool to foster long-term development processes within a touristic destination, through the introduction of entrepreneurial capacity building. The Theoretical implications show how capacity building (CB) is an essential tool to reduce poverty, to develop individual and institutional skills (UNDP, 2010) as well as a "Partnership accelerator" for effective multi-stakeholder partnership SDG 17 (UN 2030 Agenda sitography); in WHS CB is seen as a form of people-centred change that introduces and improves a dynamic relationship between the management of cultural heritage and its society (UNESCO, ICCROM, ICOMOS and IUCN, 2013). This article enhances the strong correlation between the HM approach and community capacity building.

Unfortunately, academia affirms the lack of specific studies focused on: CB applied to the tourism sector (Moscardo, 2008), specific training for entrepreneurs in WHS destinations, on cooperation as essential aspects for a business to reach a common good (Melé, 2016).

This paper gives an innovative contribution to the academic discussion by illustrating a new entrepreneurial capacity-building model, which includes the HM approach applied to community-based capacity building for entrepreneurs which enhances WHS heritage. The chapter describes the new approach both in terms of capacity-building session and overall strategy wants to make a valuable contribution to fill a gap within the scientific literature on the subject, providing a practical application to the theoretical requests.

It brings to the discussion several innovative Practical Implications, for academia, practitioners, WHS and 2030 Agenda for Sustainable Development Managers and Master's students seen as future Leaders of the tourism sector.

The "Win-Win UNESCO Experience" is a new training method designed by the author and activated by the Institutions with a "top-down" approach and managed by the trainer through "bottom-up" methodologies. Its primary objective is to make private sector entrepreneurs have a peer-to-peer dialogue with the public sector and World Heritage Site managers in order to deepen their mutual understanding and the development of lasting alliances. The capacity building introduces processes of social, business, and relational innovation, right up to the generation of the spontaneous creation of an experiential tourist product for the destination in question.

The "Win-Win UNESCO Experience" training model generates sustainable-profit tourism products for the three main demographic components of the destination system: the residents (in particular its entrepreneurs), the managers of the cultural heritage (tangible and intangible heritage) and tourists, who converse with each other in a productive way which is far from the patterns generated by mass tourism.

Each of these three players wins: as the local entrepreneurs reinvent a social role for themselves, tied to the dissemination of the intangible heritage and to the ancient know-how connected to the World Heritage Sites (WHS), the tourists can live authentic experiences linked to local entrepreneurs within the destination, while the managers of the cultural institutions can benefit from the cultural promotion carried out by the local entrepreneurs.

The product-experience created during the capacity-building sessions is sustainable from a social and economic point of view, it is durable and able to generate the resources to maintain itself in activity even many years after the training (Basile 2018a). The success is particularly evident if the model is applied to launch low seasons of popular destinations or to relaunch the marginal destinations, which have difficulties in attracting visitors all year round.

However, over the years that the "Win-Win UNESCO Experience" model has been successfully applied to various cultural, natural, and landscape WHS, some limitations to the applicability and transferability of the model have actually been observed and it is necessary to take them into serious consideration. The most frequent ones categorized by type of stakeholders are referring to: (1) Institution, which doesn't active the "top-down" phase; (2) Businesses and Entrepreneurs, which don't have sufficient internal motivation to participate continuously in all the capacity buildings, missing the opportunity to develop effective relationships with other enterprises; (3) Trainer who does not have the tools to trigger the innovative process among the entrepreneurs which is what creates the long-lasting product.

The "Win-Win UNESCO Experience" furthermore introduces an effective sustainable tourism model that could be implemented in achieving the SDGs with Goal 17, by fostering inclusive PPPs through capacity building.

The "Win-Win UNESCO Experience" training method was also introduced by the author as a guest lecturer during several training sessions of the University Master's Degrees of UNESCO Chair in order to provide students with a theoretical framework together with an immediately applicable methodology and to create with local entrepreneurs a sustainable and profitable product-experience, enhancing the World Heritage Sites.

For the second year, the author also applied the "Win-Win UNESCO Experience" capacity-building method to the University Paris 1 Panthéon-Sorbonne-IREST, UNESCO Chair during a cycle of seminars for two Master classes, one of Heritage Management, and the second one of International Tourism. Students had the opportunity to act in a real "role play" by representing the two different points of view of Cultural Institutions and Entrepreneurs in a destination where the "Win-win UNESCO experience" training model has already been activated. They had the opportunity to effectively apply the theory they learned by operating in small working groups on the creation of win-win public-private partnerships and on the product-tourism experience. During the classroom discussions and debates over the results achieved by each working group, the effectiveness of the training method is always met: the processes of creativity and flexibility coming from the students in most cases comes to the same constructive solutions elaborated by the groups of entrepreneurs in the same destination, with consequent satisfaction expressed by the students at the end of the training (University sitography).

By expanding the horizon, we could even hypothesize the introduction of this training method to the syllabus of the master's degrees of the universities that are included in the networks of international institutions that invest in durable local development through the involvement of entrepreneurs (for example, UNESCO, UNWTO, EU, OECD, Council of Europe). This would help to train a future generation of managers for the tourism sector, based on the principles of Humanistic Management that would generate significant social, territorial, and economic benefits. It would also significantly facilitate the dialogue and the construction of relationships between the different stakeholders of any given destination, reaching the SDG 17, reducing the gap between the public and private sectors. It would also be possible to create unique and sustainable product-experiences, starting from UNESCO's tangible and intangible culture, including local entrepreneurs within the decision-making processes, with application skills immediately expendable in favor of sustainable innovation processes focused on small and medium-sized enterprises.

Note

1 The author in the article indicates (WHS, year) next to the name of the World Heritage Site also the year in which it has been listed.

References

Aref, F. & Redzun, B. (2008). Tourism and Community Capacity Building: A Literature Review. *Pakistan Journal for Social Sciences*, 5(8) 806–812.

Basile, M. (2009a). Con I voli low cost nuovi clienti per gli hotel. *Turismo d'Italia*, *92*, 28–35. Milano: BE-MA Editrice.

Basile, M. (2018a). Il rilancio della destinazione Valtellina a supporto della competitività territoriale. In Becheri, E., Micera, R. & Morvillo, A. (Eds.), *Rapporto sul Turismo Italiano – XXI edizione 2016/2017* (pp. 413–425). Napoli: Rogiosi Editore.

Basile, M. (2018b). Local Food, Wine Heritage and Destination Marketing – Relaunching Valtellina Alpine Destination. *Economia della Cultura 1-2* (pp. 213–220). Bologna: Il Mulino.

Basile, M. (2019). Il metodo "win-win" applicato al prodotto turistico esperienziale. *Obiettivo Impresa*. Perugia: Camera di Commercio di Perugia.

Cipolla, C. (1996). *Teoria della metodologia sociologica – una metodologia integrata per la ricerca sociale*. Milano: Franco Angeli.

Crouch, G. I. (2011). Destination Competitiveness: an analysis of determinant attributes. *Journal of Travel Research*, *50*(1), 27–45.

Jamal, T. & Getz, D. (1995). Collaboration theory and community tourism planning. *Annals of Tourism Research* *22*(1), 186–204. USA Elsevier Science Ltd.

Kostera, M. & Pirson, M. (2017). *Dignity and Organization*. Basingstoke: MacMillan.

Loza, J. (2004). Business-Community Partnership: The Case for Community Organization Capacity Building. *Journal of Business Ethics*, *53*, 297–311.

Maclellan-Wright, F., Anderson, D., Barber, S., Smith, N., Cantin, B., Felix, R., et al., (2007). The development of measures of community capacity for community-based funding programs in Canada. *Health Promotion International*, 22(4), 299–306.

Mariani, M. M., Baggio, R., Buhalis, D. & Longhi C. (Eds.) (2014) (2015), *Tourism Management, Marketing and Development – Volume I: the importance of Networks and ICTS*. New York: MacMillan.

Melé, D. (2003). The Challenge of Humanistic Management. *Journal of Business Ethics*, *44*, 77–88.

Melé, D. (2016). Understanding Humanistic Management. *Humanistic Management Journal*, *1*, 33–55.

Moscardo, G. (2008). *Building community capacity for tourism development*. Oxfordshire: Cabi International.

OECD (2009). *The Impact of Culture on Tourism*. Paris: OECD Publishing.

Pirson, M. & v. Kimakowitz, E. (2014). Towards a human-centered theory and practice of the firm. Presenting the humanistic paradigm of business and management. *Journal of Management for Global Sustainability* 2(1), 17–48.

Pirson, M. (2017). *Humanistic Management: Protecting Dignity and Promoting Well-Being*. Cambridge: Cambridge University Press.

Richards, G. (1996) *Cultural Tourism in Europe*. Wallingford: CAB International.

Richards, G. & Raymond, C. (2000). *Creative tourism*, *23*. Wallingford: ATLAS.

Richards, G., Wilson, J. (2007). Tourism, Creativity and Development. *Series: Contemporary geographies of leisure, tourism and mobility, Book 10*. London: Routledge.

Scheyvens, R. (1999). Ecotourism and the empowerment of local communities. *Tourism Management*, *20*, 245–249.

Scheyvens, R., Banks, G. & Hughes, E. (2016) The private sector and the SDGs: The need to move beyond 'business-as-usual'. *Sustainable Development*, *24*(6), 371–382.

Smyth, J. (2009). Critically engaged community capacity building and the 'community organizing' approach in disadvantaged contexts. *Critical Studies in Education*, *50*(1), 9–22.

Tiwari, R., Lommerse, M., Smith, D. (2014). *M2 Models and Methodologies for Community Engagement* (pp. 261). Cham: Springer.

UNDP (2010). *UNDP Report on measuring Capacity*. (June 2010).

UNESCO, ICCROM, ICOMOS and IUCN (2013). *Managing cultural World Heritage – World Heritage resource manual*. Paris.

UNWTO (2015a). *Global Report on Cultural Routes and Itineraries – Affiliate members report*. Madrid.

UNWTO (2015b). *Public – Private Partnership: Tourism development*, p. 12–69 Madrid.

UNWTO (2018). *Tourism for Development, Vol.1: Key Areas for Actions*, p. 14. Madrid.

UNWTO (2019). *Guidelines for Institutional Strengthening of Destination Management Organisation (DMOs) – Affiliate members report*. Madrid.

Valentino, P. A. (2001). *I distretti culturali: nuove opportunità di sviluppo del territorio*. Civita. Roma.

Valentino, P. A. (2013). L'impresa culturale e creativa: verso una definizione condivisa, in *Economia della Cultura, Rivista trimestrale dell'Associazione per l'Economia della Cultura"*, *3*, 273–288. Il Mulino Bologna.

Vanhove, N. (1997). Mass tourism: benefits and costs. In Wahab, S. & Pigram, J. J., *Tourism Development and Growth*. London: Routledge.

Vasile, V. (2019), *Caring and Sharing: the cultural heritage environment as an Agent for Change – 2016 Alector Conference, Istanbul, Turkey*. Cham: Springer Edition.

Vidal González, M. (2008). Intangible heritage tourism and identity. *Tourism Management*, *29*(4), 807–810.

Sitography

Chamber of commerce of Perugia, https://www.obiettivoimpresaweb.it/ita/marketing_12/il-metodo-win-win-applicato al-prodotto-turistico-esperenziale_1146.html#.XkVtFGhKj64.

UN 2030 Agenda https://sustainabledevelopment.un.org/topics/capacity-building.

University Paris 1 Panthéon-Sorbonne https://www.facebook.com/IREST.Paris1/posts/2623112721115126.

10 The Humanistic Supervisor

Michael D. Santonino III

1 Introduction

This chapter examines the role of a humanistic supervisor in protecting employees' dignity. It highlights the managerial implications for humanistic supervision and outlines the best practices within the tourism industry. A humanistic tourism supervisor is sociable, community-aware, and well-informed about promoting holistic human well-being. This chapter outlines two case studies, which are examples of successful humanistic management within the tourism industry in Florence (Italy) and Singapore. Using the grounded theory to develop a framework that contextualizes the two cases, I conducted informal interviews with residents to understand the challenges involved in transforming tourism managers into humanistic supervisors. In the comparative case study, I employ a supervisory relationship framework to compare grassroots initiatives to a sustainable center of cultural education and tourism. Scholars have proposed field trips, outreach programs, change-makers, and other social-oriented roles (Pirson, 2017). Yet, many institutions struggle with the relevance of their curriculum toward social benefits. This case study's novelty incorporates the richness of images and informal chats with local residents to better understand the challenges of humanistic management in the tourism industry – an insightful approach to extending humanistic management in practice ideas (Kimakowitz, Spitzeck, Pirson, Dierksmeier, & Amann, 2011). The resulting taxonomy identifies common elements within both Singapore's and Florence's tourism industries, including a shared interest for the common good in tourism employment and a desire to educate people about sustainability in ways that might help them to change their behaviors and work for the common good to benefit society, businesses, and our planet. These are the central priorities that determine the effectiveness of the humanistic supervisor and mitigate the challenges of humanistic management.

Humanistic management is a people-oriented management approach that seeks profits in service to human ends. Notably, it contrasts with other types of management, which are essentially oriented toward profit and see people as mere resources to serve this goal (Melé, 2016). Hence, the humanistic supervisor is a relationship-oriented leader with a natural cognition of their subordinates' well-being.

Goldstein (1939), Maslow (1943), and Mayo 1933, 1946 have conducted seminal work in human behavior (including human needs and motivations) and outlined many fundamental elements of humanistic management. Furthermore, humanistic management has gained momentum in the last 10 years, resulting in several studies and discussions by scholars, practitioners, and business consultants (Kimakowitz et al., 2011; Melé, 2016; Pirson & Turnbull, 2011). Most research on humanistic management aligns with the Principles for Responsible Management Education (PRME), a tool that the United Nations Global Compact for Sustainable Development Goals (SDGs) developed in 2007 (UN Global Compact, n.d). The UN World Tourism Organization defines *sustainable tourism* as tourism that considers its current and future economic, social, and environmental impacts and addresses the needs of visitors, industries, the environment, and host communities (Hardasmalani, 2016). Notably, PRME and SDG are interrelated initiatives that address critical issues related to global social responsibility and sustainability in tourism and other industries. Key stakeholders (i.e., higher education institutions) are involved in the development of current and future managers; thus, their concerns are paramount to understanding the managerial roles and responsibilities of a humanistic supervisor. Hence, an extensive literature review is unnecessary in this context; rather, the chapter discusses some previous contributions on the interrelations between humanistic management and the SDG, as they closely align to a common theme (uniting all global stakeholders to reduce poverty, inequality, and injustice, and protect our planet).

Given that the tourism industry is inherently economy-driven, it may be difficult to shift the industry to a more humanistic view. Hence, this chapter focuses on the challenges facing a tourism supervisor, manager, or business owner and offers strategies for implementing humanistic behaviors within this industry. Doing so is of critical importance in achieving SDG as a roadmap for peace, dignity, and prosperity within the tourism industry (Winchenbacha, Hannab, & Millera, 2019). Previous work from the tourism management literature identified supervisors' incivility and self-efficacy as serious issues in organizations (Uju, Turgay, & Ozturen, 2018). Some researchers pointed out the necessity of employees' training and its effect on both the employee and the organization (Bowers & Martin, 2007; Eisingerich & Bell, 2008). This study attempts to close the gap in supervisor training by exploring the cases of Florence (Figure 10.1) and Singapore and their challenging cultural differences.

1.1 Definition of a Humanistic Supervisor

A supervisor with good human relational skills who can communicate effectively can develop genuine and trusting relationships that sustain a

Figure 10.1 The city of Florence, Italy.
Source: © Photo by Michael D. Santonino III/Kim Pei Lu Travel Photography.

department and motivate workers to go above and beyond (Santonino & Villa, 2017). This type of effective manager may be understood through the following metaphor: "A manager who is able to crawl around the production manufacturing floor and get dirt under their fingernails" (Santonino & Villa, 2017, p.4). This metaphor emphasizes the manager's ability to know what is going on in a people-oriented manner, where the word "manager" means an individual with supervisory roles and responsibilities. In this chapter, the terms "manager" and "singular" are used as synonyms. Supervisors must be aware that success is dependent on the people with whom they work; and thus, they must promote a "community of persons" – a term that traces back to the social welfare movement of the late 1800s – that works for the benefit of the whole (Melé 2003, 2012).

Organizational pluralism is dramatically changing the landscape of business and organizations, making it difficult for supervisors to know what is going on within and outside of an organization. A humanistic supervisor should consider employees as key stakeholders and show genuine concern about them beyond the employment (supervisor-subordinate) relationship. Employees' work creates value, which manifests itself throughout a business, and supervisors define the heart and soul of an organization through their leadership (Santonino & Villa, 2017), setting by example, the workplace environment, and organizational culture.

The sustainable tourism industry contains the key elements and values to build a human-centric workplace and culture. For example, the United Nations Educational, Scientific, and Cultural Organization (UNESCO) has provided a framework for tourism businesses that encourages a holistic management approach and serves as a model of democracy by encouraging different groups to voice their opinions throughout the organization. Such an organizational structure gives employees diffused power and allows divergent perspectives to converge on substantive issues affecting tourism as well as local residents and businesses. These methods have been shown to be successful, as seen through testimonials from Florence's supervisor of tourism, UNESCO supervisor, and others who encourage a holistic approach to tourism management (Korey, 2017, September 12).

However, despite the potential of tourism to incorporate humanistic management, many tourism industries around the world instead operate through an economistic approach that only benefits key stakeholders, and/or that focuses on economic growth. In these settings, it is difficult to implement a humanistic management approach. The Humanistic Management Network examined 19 examples of companies around the globe, exploring the feasibility of implementing humanistic management practices at each (Kimakowitz et al., 2011). This chapter further extends its work within the specific industry of tourism. To this end, it focuses on tourism within Singapore and Florence, Italy, outlining the challenges facing the tourism business and the best practices for achieving both social sustainability and financial success.

2 Methods

I conducted observational field trips in Singapore and Florence, both UNESCO heritage sites, and collected data using spontaneous conversations and unstructured informal interviews (informed consent was not required for either the public photos or the informal conversations). The main purpose of these case studies was to offer practical insights into the challenges facing humanistic supervisors and to explore best practices for sustainable tourism management. At the heart of sustainable tourism are people, culture, and education, which transcend an organization's holistic management approach. Over a period of five years, I observed, photographed, and interviewed local residents and tourists in Singapore and Florence. The images throughout the chapter provide a practical perspective to understand the supervisory role and practices of a humanistic supervisor within tourism management. My informal chats occurred with 10 local residents in Florence, who were holding signs listing the names of arriving guests as they waited at the Florence airport's arrival gate exit point. I selected these two countries during my work abroad trips from 2014 to 2019 as a fully immersed resident. Unstructured qualitative research interviews require adequately skilled and experienced researchers (Sorsa, Kiikkala, & Astedt-Kurki, 2014), although sensitive researchers interested in hearing stories and empathizing are beneficial (Corbin & Morse, 2003). My images, chats, secondary data, and personal reflections as a resident were analyzed using abductive reasoning to help form logical inferences from the field trip observations and offer possible explanations for the observations. Abduction is not restricted to or associated with any particular methodology. Nevertheless, the peculiarly fallible nature of abduction must be acknowledged (Lipscomb, 2012). Therefore, tabular data and narratives are integrated for each presented case, considering that case study investigators should make explicit and retrievable their 'notes, documents, tabular materials, and narratives' (Yin, 1994). This advice should not be taken as a hint to fully edit all the material collected during the research process (Swanborn, 2010).

Testimonials from secondary sources provided additional data for Florence. Informed consent was verbally given by the respondents during the factory tour in Singapore and a safety waiver provided to the participants for any potential personal incidences during travel and the factory tour. As a good practice, the researchers should be honest with the potential respondents and tell them their intentions to capture some basic information for use (Evans & Rooney, 2011). This creates a casual chat environment and a secure place for the respondents. Having people talk freely about perceived obstacles and barriers is one of the most useful assets of field studies (Swanborn, 2010). In Florence, it was made clear by the researcher that the content of the conversation would not be used for tax collection, government intervention, or by other formal authority. Finally, the images collected are all property rights of the researcher/photographer with special reprint permission courtesy of the Baseball Hall of Fame in Cooperstown New York for the 1889 image.

Patton (2002) argues that unstructured interviews are a natural extension of participant observation because they rely entirely on the spontaneous generation of questions within the natural flow of interaction. In addition, previous studies have pointed out that interviews are a complex social and organizational phenomenon rather than just a research method (Qu & Dumay, 2011). Unlike the highly structured survey questionnaires and interviews used in epidemiology and found in many highly ranked scientific journal rankings, I relied on less structured, informal interview strategies, in which the person interviewed is more a participant in meaning-making than a conduit from which information is retrieved (DiCicco-Bloom & Crabtree, 2006).

To this end, I developed a framework called the humanistic supervisor spider-web of integration to provide a guide to human relations and communication constructs in a changing worldview with an emphasis on the humanistic elements. This is an essential argument to support Pirson's theory of change, which makes a case for humanistic values, strategies, and leadership in organizations. An approach to economics and management that differentiates between the humanistic (mainly about dignity and common good for all humans) and the "economistic" approach (primarily about money, status, and wealth) (Pirson, 2017). Each element shapes tourism's humanistic perspective through a comprehensive communication network and reflects the connectivity of all stakeholders a humanistic supervisor must incorporate into tourism management. I then applied this framework to my fieldwork in each of the two case studies: Florence and Singapore.

3 Tourism in Florence, Italy

Florence is on the UNESCO Heritage Site List for its role in the development of humanism during the Renaissance. During this time, civic humanists (e.g., Leonardo Bruni) emphasized the connection between the individual and the community (known as "communitarianism"). Today, the increasing number of tourists is causing problems within the city. Figures 10.2 to 10.4 show the huge crowds that often flood the city. As noted by Anatole-Gabriel and Drury (2017), city tourism managers and government officials struggle to balance the interests of residents, visitors, and local businesses. There are increasing complaints of tourists' disrespectful behavior, including rudeness, group picnics on church steps or at the foot of historic statutes, and the disobeying of signs.

Figure 10.5 shows an example of the city's increased graffiti in local neighborhoods. A volunteer organization called Angeli del Bello (Angeli del Bello Foundation, n.d.) is working to target areas to repaint in an attempt to reclaim Florence's historic character. A volunteer says that "youth enter the city at night to degrade buildings with graffiti for kicks. Angeli del Bello provides residents a way to take back our city from the unruly". The translation of "Angeli del Bello" is "Beautiful Angels".

This grassroots, resident-led project influences behavioral change and exemplifies respect for Florence. Similarly, a city-wide campaign called "Enjoy-Respect-Florence" uses signage and multimedia efforts to educate tourists about appropriate behavior in Florence, such as basic behavioral norms designed to preserve the city's cleanliness and to respect its monuments and residents (Concia, 2017). The violation of these norms is punishable by Italian law, but the Italians prefer a softer and more welcoming action-oriented approach that focuses on educating rather than punishing tourists. Later in the chapter, I discuss Singapore's reserve action-oriented position. With the support of the minister of culture, Florence was one of five cities in Italy to incorporate such an initiative. It seeks to educate the public on the city's "good practices," which often align with the Florentine culture. The campaign also makes tourists aware of sanctions for bad behavior, which can be up to €500.

Figure 10.2 Florence police patrol the tourist zone around the Cathedral of St. Mary of Flower.
Source: © Photo by Michael D. Santonino III/Kim Pei Lu Travel Photography.

3.1 Florence as a Center Of Cultural Education and Tourism

Florence is a unique center of international cultural scholarship and has a rich history as both a place of learning and a tourist destination. The guilds and apprentice system evident in medieval times gave way to the grand tourists and collectors of the 19th century, who were, in turn, followed by today's expatriate artists and resident aristocrats. These exclusive foreign communities gave rise to many of the city's research institutes and local philanthropy behind the first restoration and conservation efforts.

Today, the city continues to have a dual appeal as both a destination and a campus, a relationship that began with the 1966 flood, when many students volunteered as "mud angels" to help the flood victims and contribute to the preservation of the historical art material (Du Terroil & Santonino, 2012). The city attracts a disproportionate number of North American students compared to other European student exchange programs (NAFSA, n.d.). Courses in subjects rooted in Florentine tradition – such as the arts, architecture, design, language, and literature – have been joined over the last 10 years by course offerings in business, fashion, and food culture.

Figure 10.3 Crowds of tourists waiting in line to climb the 450 steps to the roof of the Duomo.

This diversification is a strategic move to attract student tourists, who otherwise would not venture out of their country of origin. The strategy has been a success; today Florence is a hub for North American university programs. This has had a positive economic impact on the city, as students, faculty members, and institutes spend money throughout the year. Many of these programs have integrated tours and other local activities into their curricula. However, tourist managers remain isolated within each institution; unfortunately, there is little information exchange or communication between city tourism managers, government officials, UNESCO, Angeli del Bello, or other entities.

3.2 Developing a Humanistic Tourism Framework in Florence

Figure 10.6 shows a proposed model of tourism for study-abroad programs that would offer opportunities to develop a humanistic organization and integrated process that involves all stakeholders. The interdependencies of stakeholders center around the work of humanistic manager(s)/humanistic organization(s) to complement the city-wide "Enjoy-Respect-Florence" initiative and Angeli del Bello community efforts. Notably, when discussing the holistic impact of AirBnb on

Figure 10.4 Tour guides commonly stop in front of the Duomo.

Tuscany, Florence's mayor, Dario Nardella, advised approaching "sustainability, tourism and the challenge of legality so that all the operators in this sector are treated in the same way" (The Florentine, 2016, para.3). The humanistic supervisor spiderweb of integration considers the gap between the economic and humanistic elements by developing a communication channel with all key stakeholders. To date, each stakeholder within Figure 10.6 is an independent actor. Figure 10.7 shows a sustainability campaign advertisement in downtown Florence. Human dignity, the common good, and moral conscience are all elements of humanistic management that must coincide with economic benefit in decision-making. UNESCO, Angeli del Bello, and other foundations in Florence offer hope that more grassroots and humanistic initiatives will be successful in transforming tourism within this Renaissance city.

Developing a humanistic perspective within tourism through a comprehensive network between all stakeholders extends UNESCO's holistic approach to management by operationalizing each key stakeholder (i.e., city tourism managers, outreach organizations, local businesses, students, tourists, residents, travel agencies, and other tourism support services organizations). Awareness campaigns (e.g., social media, print ads, and signage) showed some limited evidence of success; however, the humanistic management model proposed here includes motivating

Figure 10.5 Graffiti on a train station's resident walkway, just outside of downtown Florence.
Source: © Photo by Michael D. Santonino III/Kim Pei Lu Travel Photography.

people to change behaviors to work toward a common good. In other words, organizational cultures include a community of persons who are bonded by a shared interest for the common good.

Research suggests that humanistic management skills are lacking in higher education, a problem that must be remedied to foster human dignity within a business context (Spitzeck, 2011). Training future tourism managers in Florence requires the agents of the change within organizations to balance the need for a sustainable city with the needs of cultural education and tourism industries. The humanistic supervisor spiderweb of integration can help close the communication gaps between stakeholders to prevent isolated operations. As stated by Carlo Francini, head of the UNESCO office of the City of Florence: "His office's plan provides guidelines for a holistic approach to the management of the territory and its tourism, the organization of transport and commerce, and other aspects of the city's cultural and social life. The involved entities (namely local government) have to follow this plan" (Korey, 2017, para.7).

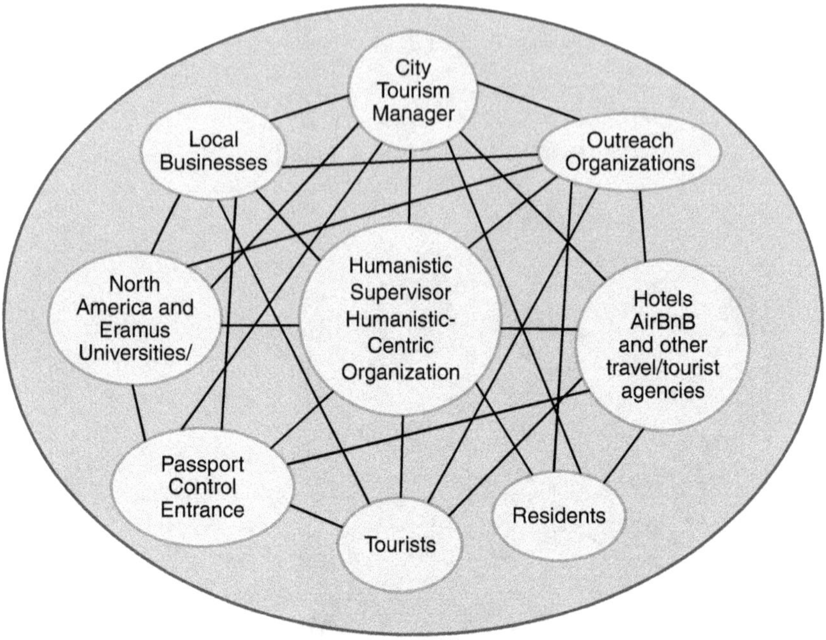

Figure 10.6 The humanistic supervisor spiderweb of integration.
Source: © Created by Michael D. Santonino III/Kim Pei Lu Travel Photography.

3.3 The History of Tourism in Florence

Tourism in Florence dates back to 1889. On February 25, Albert Goodwill Spalding, a professional U.S. baseball player and a businessman, went on a six-month world tour with his team to showcase the American sport. His Italian stop was at the Parco delle Cascine (Cascine Park) in Florence (Figure 10.8).

Spalding was forward-thinking, and his world tour exemplified humanistic leadership by encouraging people to explore their own self-transformation. This perspective was unthinkable in the late 1800s. Spalding often recited his motto to other leaders in the industry: "everything is possible for him who dares" (Lamster, 2006, 23). He also instilled a tireless leadership management style that espoused the cultural values of his business. Known as "a stickler for absolute honesty and cleanliness," Spalding managed to make his own name virtually synonymous with transparency, fair dealings, and rectitude (Lamster, 2006, 3). His humanistic management approach had an indelible social impact on Florence's tourism for over 100 years, and today this impact continues in the thousands of baseball players in Italy.

Figure 10.7 Florence allows commercial advertisements in the Piazza della Repubblica (city square).
Source: © Photo by Michael D. Santonino III/Kim Pei Lu Travel Photography.

Spalding and his two teams played at Cascine Park in 1889, located just outside of downtown Florence. For the first time, local Florentines gathered to watch professional baseball. The tour began with a procession from the players' hotel to the Cathedral of St. Mary of Flowers (famously known as the Duomo and with a historic dome designed by Brunelleschi). The structure awed many of the players, who had never visited another country, but the locals "were already in the habit of deriding the ostentatious ornamentation as too 'American'" (Lamster, 2006, 197). The tour continued to the Uffizi Gallery, the Piazza della Signoria, the Ponte Vecchio, the Pitti Palace, the Casa Buonarroti, Santa

Figure 10.8 Spalding's teams in Italy.
Source: The Baseball Hall of Fame.

Croce, Santa Maria Novella, and many other local points of interest before the afternoon game.

3.4 Changing the Model of Tourism Employment in Florence

Tourism in Florence has transcended its early origins, and today the city is one of Italy's most popular destinations. In 2018, the number of visitors approached 10 million (Statista, 2019). However, there are no official guides from educational institutions outlining how the UNESCO framework can be used to encourage a holistic management approach that alters the organizational structure and enables employees with divergent perspectives to cooperate on substantive issues that affect tourism, residents, students, and businesses. This framework is necessary to allow humanistic supervisors to act as change agents, reshaping tourism in Florence.

Today, unskilled local residents have entered the tourism industry through a variety of means, including offering airport pickup and accommodations. These residents may not have knowledge of the UNESCO framework or appropriate training. According to World Tourism Organization (2019), the tourism industry has a considerable number of informal workers, the majority of whom are female and tend to earn lower wages. Such residents are purely driven by economic considerations: they seek solely to provide income for themselves and do not consider the sustainability of tourism in the city as a whole. For example, Figure 10.9 shows local residents waiting to pick up visitors at the Vespucci airport in Florence. My informal chats with 10 of these residents over a 3-hour period revealed that the majority were seeking to supplement their income as ad-hoc tour guides or by providing housing accommodations for tourists. A Florence resident reported that they were not a professional tour guide or trained professional tour manager. Another Florence resident, a retired physician, told me he was worn out from traveling 80 km every weekend to prepare his flat for a new batch of guests in downtown Florence.

Other locals benefit from the growth of ad hoc tourism businesses by offering their homes through AirBnb. In 2016, Florence reached an

Figure 10.9 Local residents hold signs as they wait for guests to arrive at the Florence airport.

Source: © Photo by Michael D. Santonino III/Kim Pei Lu Travel Photograph.

agreement with AirBnb that allowed the city to recoup the annual estimated loss of 10 million euros in taxes due to the high turnover rates of typical short stays in the city (three days or less) (The Florentine, 2016, January 27). Yet, despite this agreement, many locals have been negatively affected by the rise in informal tourism. Some have been pushed out of the city center by tourists, AirBnb properties, legal and illegal immigrants, and commercialized cultural attractions that Italians find non-traditional. Today, city administrations are trying to reverse the trends of large crowds and the prevalence of *maleducati*, an Italian adjective used to describe rude, ill-mannered tourists.

However, these problematic issues remain. Good faith measures, including social media campaigns, marketing, and advertisements, by city tourist watchdogs have had little influence on tourist behavior. Shaping a humanistic perspective in tourism is a slow process; more change agents are needed to influence decision-makers in rethinking the need for sustainable cultural education and tourism.

Humanism struggles to adapt to rational, system-based organizations, and it is often stressful for managers to develop a humanistic management approach. Developing trust, social relations, commitment, communication, and influence will help foster this transition (Brien & Smallman, 2011). However, the implementation of a management training program that incorporates these attributes has had limited success, due to the complexity of the humanistic management perspective, which is understood and received differently by employees and managers (Requena, 2003). Despite the challenges, the humanistic supervisor spiderweb of integration (Figure 10.6) offers ways to train future tourism managers in humanistic leadership. This, along with grassroots initiatives, offers hope for the development of sustainable tourism in both Florence and around the globe.

Figure 10.10 outlines the work done by Jacobson (2009) on how to foster dignity within the tourism industry. Jacobson suggests that interaction-based dignity, such as in work environments, is influenced by the positions of and relationships between the various individual and collective actors, features of the organizational setting, and the wider social order (Jacobson, 2009; Winchenbacha et al., 2019). Scholars have utilized this model to contextualize and theorize the multilayered conditions and various actors within tourism employment (Bolton & Laaser, 2013; Gallagher, 2004; Kensbock, Jennings, Bailey, & Patiar, 2016; Winchenbacha et al., 2019).

Kant (1998), for example, links the individual and societal aspects of dignity and explains that dignity is a value beyond price. His perspective, arguably, conflicts with the predominant neoliberal structure of tourism. Florence continues to balance societal changes with humanistic management best practices and economic influencers.

Florence is rich in history and has current grassroots initiatives to educate stakeholders about a sustainable future. UNESCO, Angeli de

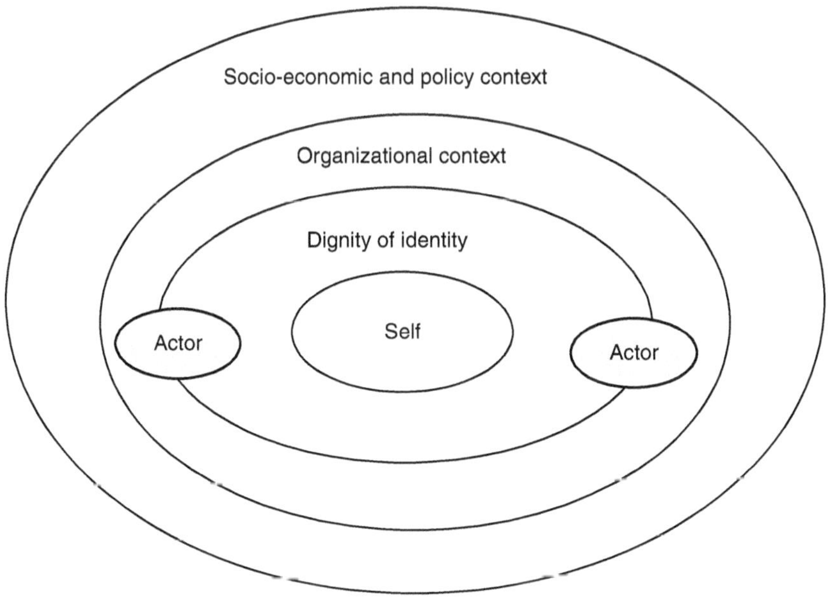

Figure 10.10 Conditions of a psychosocial understanding of dignity in tourism. Source: Jacobson (2009) and revised by Winchenbacha et al. (2019).

Bello, and other entities are attempting to improve social processes through education and awareness efforts. Today's managers are tasked with shaping human behavior and creating an environment that enables people to take initiative, cooperate, and learn from each other (Bartlett & Ghoshal, 1995). As Spalding did 131 years ago, today's grassroots initiatives in Florence allow people to explore their own self-transformation (i.e., cultural immersion via study abroad, initiatives with Angeli del Bello, Fiorentina Baseball, local business partners, and sustainable tourist activities). Such opportunities are instrumental in training future tourism managers regarding humanistic leadership.

4 Tourism in Singapore

Singapore is the most sustainable Asian country according to three subindices: people, planet, and profit (Lim, 2018. Para.4). The government is addressing several environmental challenges through social media and community initiatives, including water quality, air quality, and livability, in order to ensure a sustainable future. Changi Airport, for example, sets standards for sustainability in airport design; the airport pushes technological boundaries and maximizes operational efficiency (Lim, 2018).

Meanwhile, Marina Bay Sands also sets standards for sustainability in its emphasis on the following four priorities: green building, environmentally responsible operations, green meetings, and sustainability education and outreach (Yiannouka, 2015).

Much of the country's success is traceable to Singapore's founding father of education, former Prime Minister Lee Kuan Yew. Yew's forward-thinking vision of education has influenced the social mindset for generations. His strategy is often described as an attempt "to develop Singapore's only available natural resources, its people" (Yiannouka, 2015). Yew's humanistic management approach left an indelible mark on many Singaporeans and continues to shape social and economic humanistic-centric attributes found in humanistic leadership.

Experiential learning has successfully brought real-world community involvement in the classroom. One local university integrated a tour into its program of study in an attempt to develop community partnerships and foster opportunities for local businesses to tap into its knowledge resources. Another operation manager instructor integrated factory tours into classes to further extend the "community of persons" model and incorporate a common good approach (Figures 10.11–10.13). This method allowed students to explore local factories while providing pro bono support services to business owners, including marketing plans, logo design, productivity techniques, and other business needs. These visits were a win for both students and business owners and allowed all of them to focus on the common good: the students gained knowledge, and business owners were able to outsource their business services.

Sustainable Singapore Blueprint, a government-sponsored initiative, and Training Industry Professionals in Tourism, a professional association, both demonstrate how Singapore is planning for the future of tourism (Singapore Tourism Board, 2019; Ministry of the Environment and Water Resources and Ministry of National Development, 2015). Both organizations demonstrate SDG-aligned attributes for sustainable tourism, education, and culture. Adding more of these organizations and grassroots initiatives within the local community will begin to transform training for future tourism managers.

4.1 Singapore as a Center of Cultural Education and Tourism

Singapore is on UNESCO's intangible cultural heritage list due to its botanic gardens and food-hawker culture (Figure 10.14). However, in other ways, the country has not taken great strides toward sustainability; for example, the lack of rivers, strong winds, or landmass for large-scale solar panels restricts the nation's ability to develop a diversified energy strategy. Thus, enhancing Singapore's natural resources rather than exploiting them remains a challenge within the tourism industry. To integrate sustainable tourism with the humanistic management approach is to balance economic

Figure 10.11 Khong Guan Biscuit Factory, Singapore.
Source: © Photo by Michael D. Santonino III/Kim Pei Lu Travel Photography.

Figure 10.12 Khong Guan Biscuit Factory Museum, Singapore: tourists viewing the company's history and product offerings.
Source: © Photo by Michael D. Santonino III/Kim Pei Lu Travel Photography.

drivers (i.e., creation of wealth) with human needs and values intrinsic to employees and tourists. In other words, the tourism industry must balance an economic productivity mindset with human elements and show that it genuinely cares about its people, their needs, and their lives. Scott

Figure 10.13 Khong Guan Biscuit Factory: All tourists receive biscuits at the end of the tour.
Source: © Photo by Michael D. Santonino III/Kim Pei Lu Travel Photography.

Thompson (2016) argues that implementing such a framework is difficult due to the complexity of human behavior and ethical questions.

Similarly to Florence, Singapore would benefit from a comprehensive network connecting all stakeholders, as shown in Figure 10.6. Unlike Florence, awareness campaigns, including social media, print, and signage, are further along in Singapore, and the country has made moves toward eco-friendly building and other sustainable blueprints. For example, the Ministry of Environmental and Water Resources provides a comprehensive social media platform and field agents to educate people on how to change their behaviors to meet the common good. However, many of these initiatives are still in the early development stage and more work is needed to build awareness, develop partnerships, and establish grassroots action plans.

Educators in Singapore have introduced local factory tours to raise awareness of the UN's Sustainable Development Goals, a grassroots effort that is influencing future tourism. Skills Future, an initiative of the Singapore government, is another example of collaborative partnerships aiding future tourism supervisors. The program provides information on the tourism industry, career pathways, occupations, and job roles and offers training programs to improve skills and support future managers (Skills Future, n.d.).

Despite these attempts, the literature shows that Singapore's tourism focuses on economic gain rather than a balance between economic and human needs. This development should be observed critically; the growth of tourism and changes in the relationships between humans and the environment develop in parallel with the consumption and depletion of natural resources (Gossling, 2002).

5 Discussion

Shaping humanistic management within tourism involves changing human behavior. Both Florence and Singapore highlight the growing need for tourism supervisors to develop a more sustainable mindset transitioning from being economy-driven to humanistic. Perhaps the transition would involve both economy-driven and humanistic elements, as social entrepreneurs (i.e., change agents) can serve as templates for humanistic management (Austin, 2006; Pirson, 2009). Luckily, UNESCO and the UN Global Compact are slowly bringing about these behavioral changes.

Many sustainability and humanistic management approaches focus on changing individual decisions through awareness. The recycling of plastic waste is a good example of this. Hi Fly Airlines (2019), a low-cost airline based in Lisbon, recently became the first plastic-free airline company in the world; they removed all plastic spoons, forks, and knives, cups, serving trays, water bottles, and so on, and replaced these cabin service items with paper or other bio-friendly alternatives. In this and other examples,

Figure 10.14 Over 100 food-hawker centers are located throughout Singapore. Source: © Photo by Michael D. Santonino III/Kim Pei Lu Travel Photography.

managers set the workplace environment and organizational culture and employees will follow their lead. Florence and Singapore are both recognized by UNESCO, which encourages changing behaviors by working in the community toward a holistic management approach. For example, Singapore's health agencies are sending out field ambassadors to educate people on healthier habits, including nutrition (Khalik, 2018). Such initiatives have a ripple effect in the tourism industry, as more businesses integrate a sustainable strategy into their business model.

6 Conclusion

If a business focuses solely on economic goals and interests, then it is difficult to implement a humanistic management approach. The training of future tourism managers should focus on getting them to believe in the organization's culture and set the tone for employees. The culture of an organization is the pulse of the people, and the supervisor is the one who sets the tone. These are the central priorities to educate and train the humanistic supervisor and mitigate the challenges of humanistic management. The observational field trips in Singapore and Florence are small practical examples of the difficulties to implement the humanistic management approach. Yet, a shared interest for the common good in tourism employment and a desire to educate people about SDGs, PRME, and other sustainability initiatives can (slowly) occur to benefit society, businesses, and our planet. Despite the challenges, the humanistic supervisor framework presented in this chapter offers ways to train future tourism managers, increase stakeholder participation, and encourage grassroots initiatives to (slowly) change the economy-driven tourism industry. However, due to the nature of inclusive stakeholders, it can be a challenge for supervisors to balance these groups fairly in decision-making. A supervisor must align his or her own awareness of the important elements of managing people and be aware that an organization can change as people change.

The future of humanistic research should focus on operationalizing the construct with the human elements, talking to people to educate them. The question of feasibility will become less relevant as more grassroots initiatives become the common good for all. Hopefully, this chapter will attract sufficient attention from researchers and inspire more humanistic supervisors.

References

Anatole-Gabriel, I. & Drury, P. (2017). *Joint UNESCO world heritage centre-ICOMOS advisory mission to the world heritage site of the historic centre of Florence, Italy*. Paris, France: UNESCO. Retrieved from http://whc.unesco.org/en/documents/160848.

Angeli del Bello Foundation (n.d). Retrieved from http://www.angelidelbello.org/.

Austin, J., (2006). Social Entrepreneurship: It's for Corporations, Too. In A. Nicholls (Ed.) *Social Entrepreneurship: New Models of Sustainable Social Change* (pp. 169–180), New York: Oxford University Press.

Bartlett, C. & Ghoshal, S., (1995). Changing the role of top management beyond systems to people. *Harvard Business Review*, *73*(3), 132.

Bolton, S. C. & Laaser, K., (2013). Work, employment and society through the lens of moral economy. *Work, Employment and Society*, *27*(3), 508–525. doi: 10.1177/0950017013479828.

Bowers, M. & Martin, C., (2007). Trading places redux: Employees as customers, customers as employees. *Journal of Services Marketing*, *21*(2), 88–98. doi: 10.1108/08876040710737859.

Brien, A. & Smallman, C., (2011). The respected manager...the organisational social capital developer. *International Journal of Hospitality Management*, *30*, 639–647. doi: 10.1016/j.ijhm.2010.11.009.

Concia, A., (2017, August 10). Interview by *The Florentine*. Retrieved from https://www.theflorentine.net/2017/08/10/enjoyrespectfirenze/.

Corbin, J. , Morse, J. (2003). The unstructured interactive interview: issues of reciprocity and risks when dealing with sensitive topics. *Qualitative Inquiry*, *9*(3), 335–354. doi: 10.1177/1077800403009003001.

DiCicco-Bloom, B. & Crabtree, B., (2006). The qualitative research interview. *Medical Education*, *40*(4), 314–321. doi: 10.1111/j.1365-2929.2006.02418.x.

Du Terroil, R. & M. D.Santonino, III. (2012). The American study abroad industry in Italy. *International Journal of Arts and Sciences*, *5*(5), 109–120.

Eisingerich, A. & Bell, S., (2008). Perceived service quality and customer trust: Does enhancing customers' service knowledge matter? *Journal of Service Research*, *10*(3), 256–268. doi: 10.1177/1094670507310769.

Evans, A. & Rooney, B., (2011). *Methods in Psychological Research*. Thousand Oaks, CA: Sage.

The Florentine (2016, January 27). Airbnb and Florence sign tourism tax agreement. *The Florentine*. Retrieved from http://www.theflorentine.net/news/2016/01/airbnb-and-florence-sign-tourism-tax-agreement/.

Gallagher, A., (2004). Dignity and respect for dignity-two key health professional values: Implications for nursing practice. *Nursing Ethics*, *11*(6), 587–599. doi: 10.1191/0969733004ne744oa.

Goldstein, K., (1939). *The organism: A holistic approach to biology derived from pathological data in man*. Salt Lake City, UT: American Book. doi: 10.1037/10021-000.

Gossling, S., (2002). Human–environmental relations with tourism. *Annals of Tourism Research*, *29*(2), 539–556. doi: 10.1016/S0160-7383(01)00069-X.

Hardasmalani, R., (2016). Higher demand for eco-tourism as environmental awareness increases. Retrieved from https://www.todayonline.com/business/higher-demand-eco-tourism-environmental awareness-increases.

Hemmerdinger, J., (2019). Boeing ups 20-year forecast on global growth projections. *FlightGlobal*. Retrieved from https://www.flightglobal.com/airframers/paris-boeing-ups-20-year-forecast-on-global-growth-projections/133117.article.

Hi Fly Airlines. (2019). New campaign to end plastic pollution. *Media Center*. Retrieved from https://www.hifly.aero/media-center/new-campaign-to-end-plastic-pollution/.

Jacobson, N., (2009). A taxonomy of dignity: A grounded theory study. *BMC International Health and Human Rights*, 9(1), 3. doi: 10.1186/1472-698X-9-3.

Kant, I., (1997). *Critique of Practical Reason*. Cambridge: Cambridge University Press. doi: 10.1017/CBO9780511770869.

Kant, I., (1998). *Groundwork for the metaphysics of morals*. Cambridge: Cambridge University Press.

Kensbock, S., Jennings, G., Bailey, J. & Patiar, A., (2016). Performing: Hotel room attendants' employment experiences. *Annals of Tourism Research*, *56*, 112–127. doi: 10.1016/j.annals.2015.11.010.

Khalik, S., (2018). Singaporeans adopting better diets, easing up on calories: Survey. *Straitstimes*. Retrieved from https://www.straitstimes.com/singapore/health/sporeans-adopting-better-diets-easing-up-on-calories-survey.

Kimakowitz, E. V., Spitzeck, H., Pirson, M., Dierksmeier, C. & Amann, W., (2011). Introducing this book and humanistic management. In E. V. Kimakowitz, H. Spitzeck, M. Pirson, C. Dierksmeier & W. Amann (Eds.) *Humanistic management in practice* (pp. 1–12), New York: Palgrave Macmillan. doi: 10.1057/9780230306585.

Korey, A., (2017, September 12). Italy vs. the tourists. The summer it came to a head. *The Florentine*. Retrieved from http://www.theflorentine.net/art-culture/2017/09/italy-tourists/.

Lipscomb, M., (2012). Abductive reasoning and qualitative research. *Nursing Philosophy*, *13*(4), 244–256. doi: 10.1111/j.1466-769X.2011.00532.x.

Lamster, M., (2006). *Spalding's world tour: The epic adventure that took baseball around the globe*. New York: PublicAffairs.

Lim, S., (2018, October 31). Singapore is the most sustainable Asian city, 4th in the world: Arcadis. *Business Insider Singapore*. Retrieved from https://www.businessinsider.sg/singapore-is-the-most-sustainable-asian-city-4th-in-the-world-arcadis/.

Maslow, A. H., (1943). A theory of human motivation. *Psychological Review*, *50*(4), 370–396. doi:10.1037/h0054346.

Mayo, E., (1933). *The human problems of an industrial civilization*. New York: Macmillan.

Mayo, E., (1946). *The social problems of an industrial civilization*. Cambridge: Cambridge University Press.

McMahon, B., (2016). Al Spalding. Society for American Baseball Research. Retrieved from https://sabr.org/bioproj/person/b99355e0.

Melé, D., (2003). The challenge of humanistic management. *Journal of Business Ethics*, *44*(44), 77–88.

Melé, D., (2012). The firm as a "community of persons": A pillar of humanistic business ethos. *Journal of Business Ethics*, *106*(1), 89–101. doi: 10.1007/s10551-011-1051-2.

Melé, D., (2016). Understanding humanistic management. *Humanistic Management Journal*, *1*(1), 33–55. doi: 10.1007/s41463-016-0011-5.

Ministry of the Environment and Water Resources and Ministry of National Development. (2015). Retrieved from https://www.mewr.gov.sg/ssb.

National Association of Foreign Student Advisers (NAFSA, n.d.). Retrieved from https://www.nafsa.org/policy-and-advocacy/policy-resources/trends-us-study-abroad.

Patton, M. Q., (2002). *Qualitative research and evaluation methods*. Thousand Oaks, CA: Sage.

Pirson, M., (2017). *Humanistic Management: Protecting Dignity and Promoting Well-Being*. Cambridge: Cambridge University Press.

Pirson, M., (2009). Social Entrepreneurship:A Blueprint for Humane Organizations? In H. Spitzeck, M. Pirson, W. Amann, S. Khan & E. von Kimakowitz (Eds.), *Humanism in Business* (pp. 248–259). Cambridge: Cambridge University Press.

Pirson, M. A. & Turnbull, S., (2011). Towards humanistic governance. *Journal of Business Ethics*, *99*(1), 101–114. doi: 10.1007/s10551-011-0752-x.

Qu, S. & Dumay, J., (2011). The qualitative research interview. *Qualitative Research in Accounting & Management*, *8*(3), 238–264. doi: 10.1108/11766091111162070.

Requena, F., (2003). Social capital, satisfaction and quality of life in the workplace. *Social Indicators Research*, *61*, 331–360.

Santonino, M. D. & Villa, F. J. (2017). *The supervisor's guide to human relations & communication*. Scotts Valley, CA: Create-Space Independent Publishing Platform.

Singapore Tourism Board. (2019). Training industry professionals in tourism (TIP-iT). Retrieved from https://www.stb.gov.sg/content/stb/en/assistance-and-licensing/grants-overview/training-industry-professionals-in-tourism-tip-it.html.

Skills Future (n.d). Retrieved from https://www.skillsfuture.sg/skills-framework/tourism.

Sorsa, M., Kiikkala, I., Astedt-Kurki, P., (2014). Bracketing as a skill in conducting unstructured qualitative interviews. *Nurse Researcher*, *22*(4), 1–4. doi: 10.7748/nr.22.4.8.e1317.

Spitzeck (2011). An integrated model of humanistic management. *Journal of Business Ethics*, *99*, 51–62.

Statista (2019). Retrieved from https://www.statista.com/statistics/722442/number-of-tourists-in-florence-italy/.

Swanborn, S., (2010). *Case Study Research: What, Why and How?* Sage publications.

Thompson, S., (2016). Challenges of humanistic management. Retrieved from http://humanisticmanagement.org/cgi-bin/adframe/knowledge_center/journalistic_and_miscellaneous/article.html?ADFRAME_MCMS_ID=247.

Uju, V. , Turgay, A. & Ozturen, A., (2018). Organization Sustainability through Human Resource Capital: The Impacts of Supervisor Incivility and Self-Efficacy. *Sustainability*, *10*(8), 1–16. doi: 10.3390/su10082610.

United Nations Educational, Scientific and Cultural Organization (UNESCO) Heritage Site. (2017). *Joint UNESCO world heritage centre-ICOMOS advisory mission to the world heritage site of the historic centre of Florence, Italy*

(Publication prepared by Anatole-Gabriel & Drury). Retrieved from http://whc.unesco.org/en/documents/160848.

UN Global Compact (n.d). Retrieved from https://www.unglobalcompact.org/take-action/action/management-education.

UNWTO (2019). International tourist arrivals reach 14 billion two years ahead of forecasts. *World Tourism Organization*. Retrieved from https://www.unwto.org/global/press-release/2019-01-21/international-tourist-arrivals-reach-14-billion-two-years-ahead-forecasts.

Winchenbacha, A., Hannab, P. & Millera, G., (2019). Rethinking decent work: The value of dignity in tourism employment. *Journal of Sustainable Tourism*, 27(7), 1026–1043. doi: 10.1080/09669582.2019.1566346.

Yiannouka, S., (2015). The secret of Singapore's success in education. *Straitstimes*. Retrieved from https://www.straitstimes.com/opinion/the-secret-of-singapores-success-in-education.

Yin, R. K., (1994). *Case study research: design and methods*. Thousand Oaks, CA: Sage.

11 Natural History Museums and Sustainable Development

The Role of Education for Humanistic Tourism

Stefania Oliva and Luciana Lazzeretti

1 Introduction

Since the Brundtland Commission introduced the concept of "sustainable development" (WCED, 1987), it has increasingly characterized the models of growth of different economic sectors. Sustainability has become a fundamental theme for the development of cities, where the problems caused by globalization, climate change, and unsustainable urbanization are increasingly pressing. This has led to the attempt to develop policies, practices, technologies, and innovations for more sustainable management of cities (Camagni, Capello, & Nijkamp, 1998; Bulkeley & Betsill, 2005; Ahern, 2011; Caloffi, Pryke, Sedita, & Siemiatycki, 2017).

In particular, the management of cultural heritage has been the object of this growing sensitivity, concerning issues related to the preservation of cultural and environmental resources. In a policy perspective, the European Council (2014) recognizes that "*cultural heritage plays a specific role in achieving the Europe 2020 strategy goals for a "smart, sustainable and inclusive growth" because it has social and economic impact and contributes to environmental sustainability*" (p. 2). This has been accompanied by an increasing focus on sustainable tourism practices, where the relationship between tourism, communities, and nature has become central (Della Lucia & Franch, 2017). Among the goals of the "Agenda 2030 for Sustainable Development", two objectives concern the efforts to protect and safeguard the world's cultural and natural heritage and promote policies and practices for sustainable tourism (United Nations, 2015).

In this context, the role of science museums is fundamental, not only because they represent the temples for the conservation of historical and naturalistic collections, but also because they are the places of study and research on the topics related to sustainability and ecology. Although their function is to preserve environmental and cultural resources and enhance collections for disseminating knowledge of safeguarding cultural and natural sites, the role of museums for promoting programs of sustainable development is a still under-researched topic (Pencarelli, Cerquetti, & Splendiani, 2016).

This preliminary research wants to enlarge the debate on the role of museums in fostering the dissemination of knowledge of sustainable development. Through innovative and interacting practices, research activity, and publications, museums may represent informal spaces where visitors, the scientific community, and society may learn the importance of nature and biodiversity for sustainable development. The analysis focuses on the case of the Natural History Museums (NHM) of Florence, analyzing the activities through which it disseminates knowledge of environmental issues. Exploring the activities developed in education, research, and scientific publications, the article aims to understand how the NHM of Florence may contribute to create and disseminate knowledge of science, environment, and biodiversity. The analysis wants to respond to the following research question: "How do naturalistic museums may contribute to create and disseminate knowledge and make visitors aware of and participant in the issue of sustainability?"

The chapter is structured as follows. Section 2 presents a theoretical debate concerning the environmental threats due to the growth of the tourism sector and the need for shaping tourism practices through a more humanistic vision. Moreover, it discusses the importance of education for sustainable development and the role of museums. Section 3 addresses the research design debating the methodology applied, tools for data collection, and introduces the case study. In Section 4, the case of NHM of Florence is discussed focusing on the activities analyzed in terms of dissemination of knowledge. Finally, Section 5 is devoted to discussing the results of the analysis and drawing conclusions for the management of museums.

2 Theoretical Background

2.1 Tourism and Environmental Sustainability: Looking for a Humanistic Vision

Economic, social, and environmental development requires a strict correlation between material factors, such as physical and environmental capital and immaterial factors, such as human capital comprehending education, capabilities, and local knowledge (Sacco, Blessi, & Nuccio, 2009). This vision frames in sustainable development and implies creating ways to relate communities, institutions, and companies in order to construct a sustainable society. Sustainable development is a complex and multifaceted concept and implies to achieving human development in an inclusive, connected, equitable, prudent, and secure manner (Gladwin, Kennelly, & Krause, 1995). The three faces of sustainability – economic, social, and environmental – are distinct but overlap as social and economic sustainability are prerequisites for environmental sustainability (Goodland, 1995). Moreover, economic performances may deeply influence the social and environmental aspects. This entails that the totality of economic sectors should seriously consider the sustainability issue, and governments should develop practices for sustainable development based on the companies' impact (Elkington, 2013).

Over the last few years, the tourism sector has increased its economic impact. It is considered one of the most important industries worldwide, and, through a number of different activities, it contributes to increasing the competitiveness of many regions (Alberti & Giusti, 2012; Capone, 2016; Della Lucia & Segre, 2017). An estimation of UNWTO reports that international tourist arrivals increased 6% worldwide in 2018 (UNWTO, 2019), while the direct contribution of travel and tourism to GDP in 2018 was 3.2% (WTTC, 2019).

However, along with the growth of the economic performance, worries about its social and environmental impact have risen and how to make touristic activities more sustainable for societies (Morgan, 2004; Sharpley, 2009). The future development of the sector is facing new challenges related to economic, social, political, environmental, and technological issues. This requires the orientation of policymakers to create integrated strategies capable of involving all stakeholders in tourism governance (OECD, 2018). Tourism has a double relationship with the natural environment. Touristic activities consume environmental or man-made resources to allow visitors to know and enjoy touristic places. Moreover, the tourism sector and related industries produce externalities that have an impact on the environment (Briassoulis & Van Der Straaten, 2013). The negative impact of the tourism sector is related to several factors. Tourism increases emissions, the use of water, and energy resources. It creates waste and requires massive infrastructural development (McKercher, 2003). Uncontrolled tourism in cities of art has an impact on the carrying capacity of monuments and historical buildings located in the city centers. Moreover, an overload of tourism often means the depopulation of city centers by local communities, implicating a loss of authenticity of places (Lazzeretti, 2012).

Tourism management should take into account a human perspective in order to establish a balance between economic, social, and environmental sustainability. A humanistic management perspective looks at economic organizations as communities of persons based on common ideals and values that may positively impact individuals and society (Melé, 2003). It embraces a path devoted to improving the connection between management and social well-being (Pirson, 2014). Sustainable development may be connected to the equity of access to natural resources and in the distribution of the costs and benefits that derive from the use of these resources (Hunter, 1997).

This vision applied to the management and planning of the touristic sector emphasizes the need for a coordinated vision in the development of policies of cities and regions and to integrate a perspective of sustainability and corporate social responsibility in organizations' programs (Sharpley, 2009; Briassoulis, & Van Der Straaten, 2013). Therefore, the development of sustainable tourism in a humanistic perspective should focus on the human being by promoting individual development in accordance with a growth that takes into account social and environmental factors with the goal of producing a positive impact on the community. This requires rethinking the relationship between tourists, citizens, and tourist organizations in order to develop an offer that may contribute to creating more conscious tourism.

Within this frame, education may have a crucial role. The study of the relationship between education and sustainable development is increasing (Dettori, Caboni, & Giudici, 2017). Sen (1999) argues that human behavior is a central issue in the discussion about sustainable development and education is a crucial element for sustainability. Environmental education may aim to facilitate the adoption of sustainable practice.

2.2 *The Educative Role of Museums*

Over the last few years, culture has moved from being studied through an instrumental economic lens to being considered for its capacity of producing innovation and sharing value for society, focusing on its innovative capacity, and social quality (Sasaki, 2010; Bertacchini & Segre, 2016; Greffe, 2016). Lazzeretti (2012) discusses the societal function of cultural heritage as the capacity of culture to "share value". In this sense, material and immaterial heritage become a source for democracy and well-being and an asset for designing new forms of "sustainable cultural capitalism".

Museums are facing continuous processes of profound transformation, from "economic engines" of growth and innovation to "social engines". For many years, they have been considered the core element of urban tourism and an important resource for developing a 'high-quality' urban environment (Thyne, 2001). However, a growing literature started to consider the value produced by culture and museums and the reasons why visitors should participate in their activities (Hutter & Throsby, 2008; Scott, 2009). The value produced is not only linked to price or aesthetic but may concern the impact on individuals, such as the educative value, or on society, such as participatory forms of social inclusion and cohesion. This vision leads to rethinking the role of the museums in society, where they may be crucial for interpreting the paradigm of open culture achieving their institutional purpose (Solima, 2017).

Among the value produced, their educative function is assuming centrality in the contest of management of the cultural organization. Modern museums may represent educational and research spaces and contribute to the production of knowledge (Lazzeretti & Capone, 2015). In particular, science and natural history museums host collections that attest to biodiversity. They are the bridge between the scientific community and the public and contribute to the conservation of naturalistic collections and the promotion of education of biodiversity. Despite taxonomy research on collections rarely produces high-cited scientific literature and museums' management prefers to invest in activities for the public more than on research, it is crucial to contribute to environmental issues and to safeguard ecosystems. Through processes of cataloging of biodiversity, naturalistic museums improve the taxonomy of species and the protection of nature (Andreone et al., 2014). Moreover, today's museums and science centers provide experiences that encourage visitors to explore not only historical artifacts and scientific phenomena but also contemporary social and environmental questions, such as climate change, environmental pollution, or man-hand disasters. For this renewed function, museums may be considered an alternative channel to disseminate knowledge of sustainability (Ballantyne & Packer, 2005). It is important to note how their educative role has changed over time, moving from guided tour for schools or adult visitors to a more planned and articulated activity that involves exhibitions, workshops, and the use of digital technology. In particular, the technological transition has given to museums the opportunity to enlarge their public through the support of ICT. This embraces a number of new instruments for contents communication and development and audience engagement, going from the use of social media and mobile devices, virtual and augmented reality, and value co-creation (Cerquetti & Ferrara, 2018). The technological empowerment of museums represents a challenge for the enhancement of cultural heritage and their educative role. Through an appropriate contextualization and communication of the value of their collections, museums may overcome their physical and temporal borders in order to develop new forms of entertainment and education (Cerquetti, 2018; Solima, 2018).

This renewed context has led to the development of a new managerial approach to education and research activity. On the one hand, the new educative function requires new skills for cultural operators to reconcile the needs of a cultural offer suited to the values of the institution with the analysis of the new cultural demand and visitor flows (Hooper-Greenhill, 1999). On the other hand, due to the approach of new public management, which asks for increasing the efficiency of the cultural organization (Belfiore, 2004), higher-quality scientific research developed by museums should increase possibilities to access economic resources.

In this emerging context, the educative role of museums should carefully consider a nested idea of sustainability, where the economic, social, and environmental dimensions cannot be alternatives to one another but interact in a co-dependent way. This means that the relationship between sustainable development and museums should consider not only their attitude to disseminate sustainability issues through non-formal education and research. Also important is their role as drivers for social change, which through accessibility, relationships with local and regional authorities, may give to museums the role of promoters of good practices (Sutter & Worts, 2005; Pencarelli, Cerquetti, & Splendiani et al., 2016). In this context, the financial dimension plays a fundamental role, meaning attraction of new visitors, financial autonomy, and self-sufficient economic management. This requires managerial qualities and the ability to develop long-term relationships with the public through the continuous innovation of museums' offer (Di Pietro, Guglielmetti Mugion, Renzi, & Toni, 2014). Such considerations are contrary to the insufficient financial resources of public museums and require a careful reflection by museums management in order to maintain a balance between the different dimensions of sustainability and the role that they may play for society.

3 Research Design

The research is based on a qualitative analysis following the case study approach (Yin, 2017). Such a methodology could be useful because of the relatively unexplored nature of the research topic (Eisenhardt, 1989). The analysis focuses on the cases of the NHM of Florence and how the activities and management of the museum deal with the issues of sustainable development.

The NHM of Florence is a university museum and is part of the University Museum System, an institution devoted to conserving, enhancing, and promoting the heritage of the University of Florence. The oldest part of the museum is represented by the botanical garden "Giardino dei Semplici", commissioned in 1545 by Cosimo I of Medici. The institution of the museum dates back to 1775 when Grand Duke Pietro Leopoldo supported the creation of the first naturalistic museum opened to the public, the 'Imperial and Royal Museum of Physics and Natural History' at Torriggiani Palace. The collection, acquired by the University of Florence, was the basis to constitute the NHM that today counts eight collections in three sites: "Palazzo Nonfinito" with the ethnoanthropological and osteological collections; "La Specola" with anatomical ceroplastic collections, litho-mineralogical and zoological collections; "La Pira" with the geo-paleontological and botanical collections and the Botanical Garden. The system also counts of two historical villas, Villa la Quiete and Villa Galileo. The totality of the University Museum System staff includes 52 persons consisting of curators, maintenance staff, gardeners, photographers, archivists, and administrative service personnel.

The reason behind studying the NHM of Florence is that naturalistic museums usually pay more attention to issues related to environmental sustainability due to the nature of their collection. They enhance learning and engagement in science and are highly prolific in the production of scientific research (Mujtaba, Lawrence, Oliver, & Reiss, 2018). Moreover, for its historical importance – it conserves the largest naturalistic collection of Italy – and for the proactive strategies developed in communication and reporting activities, the NHM of Florence has been the unit of analysis of a number of scientific researches in economics and management (Manetti & Valeri, 2012; Lazzeretti, Sartori, & Innocenti, 2015; Lazzeretti & Capone, 2015). However, still, limited literature exists on the role of NHM in sustainable development.

The case study is constructed using three different methods of analysis. First, a period of participated observation has been carried out in the organization from February 2018 to June 2019. Participant observation is a methodology that allows collecting data through daily interaction with subjects (Jorgensen, 2015). Second, primary data have been collected through in-depth interviews with the different actors of the museum. In-depth interviews have been focused on collecting information about three main topics: (1) the organization and activities of the department in which the respondent operated; (2) the responsibility of museums and their role in society; (3) projects and activities with social and environmental impact; (4) contradictions in the environmental, economic, and social values, of museums. Finally, the third phase of analysis on secondary data has been carried out through the consultation of documents, reports, websites, and social media and has allowed triangulation of information (Yin, 2017). Table 11.1 shows the different sources for data collection.

The analysis of data followed three main steps: (1) describing the identity of the museum and characteristics of its visitors and stakeholders; (2) mapping of the activities developed; (3) discussing the most highly impacting activities on environmental concerns and sustainability.

Data collection involved complementary sources (Stake, 2013), and it has been held from February 2018 to June 2019:

- Participant observation: participation in meetings, conferences, and events promoted and organized by the museum.
- 10 in-depth interviews: they were conducted in person and lasted between 30 and 120 minutes.
- Collection of secondary data: museums catalogs, annual reports, research publications, general publications, webpages, social media, other documents.

Table 11.1 List of the primary and secondary sources of data collection

Primary source	
Participatory observation	
Interviews	
1	Head of the Botany Section
2	Administration Staff
3	Head of "La Pira" site
4	Head of the Administration Department
5	Head of the Zoology Section
6	Head of the Mineralogy Section
7	Head of "Palazzo Nonfinito" site
8	Director
9	Curator of Continental Molluscs Collection
10	Head of the Geology and Palaeontology Section
Secondary sources	
Social report of the University of Florence 2016, 2017	
Mission report of the University Museum System 2018	
Customer Satisfaction Analysis 2018	
Website and social media	
Other reports and publications	

Source: Authors' elaboration.

4 The case of the Natural History Museum of Florence

The relational capacity of museums has changed over time, covering the organization of a multiplicity of events and interacting with the public through new instruments of ICT, used also to communicate the results of scientific research (Lazzeretti & Capone, 2015). According to the literature review, museums may contribute to activating participant processes on the themes of sustainable development through their educational function, dissemination, and creation of knowledge.

The educational function of the NHM of Florence originates from the institution of the museum in 1775. At that time, some pieces of the collection, as the anatomical waxes, were accompanied by a series of watercolors that had an explanatory and self-taught function for visitors. Over time, the collections gave rise to classes of higher education in the natural sciences and it was split into different sites. Due to this, the museum lost its initial uniqueness and gradually took on today's structure (Barbagli & Pratesi, 2009).

In order to assess the impact of NHM on topics related to sustainable development, the analysis has contributed to identifying the number and characteristics of visitors, events, and exhibitions, the contributions to scientific research, and the research projects. Analyzing visitors, it is possible to understand the target of the public with which the museum interacts and what strategy can be successful in order to communicate to existing visitors and to attract new ones. A discussion concerning events and exhibitions may give an idea of which are the priority themes for the museum management and curators. Finally, a focus on research activities may give insights on the partners of the museum, in order to evaluate the range of the dissemination of knowledge and how they share good practices.

4.1 Visitors, Events, and Exhibitions

According to the mission report of 2018, the totality of the exhibitions of the NHM has been visited by around 108,000 persons in 2018. The most important target of the museum's public has been children between 0 and 18 years, confirming that it has a crucial role in sensitizing the new generations to the themes of natural history. According to the data collected from the customer satisfaction analysis, 53% of the public of the Geologic and Paleontologic section is represented by this target of young visitors, 50% for the Zoological section, and 39% for Anthropology and Ethnology.

Considering the target of the museum, an important role is played by the educational services. They organize a range of educational activities for all age groups in order to disseminate knowledge of nature in school education and lifelong learning. According to the mission report of 2018, educational activities include guided tours to the collections and in-depth thematic visits of 2 hours. Activities mostly involve schools and private groups or associations. In 2018, the NHM has conducted 875 guided tours and 288 thematic visits, for a total of 1,163 educational activities and an estimated number of 23,260 participants. The programs of educational activities are divided on the basis of the degree of education of participants in four main categories: kindergarten, primary school, secondary school, and high school. This allows education staff to plan specific activities based on the attitude and knowledge of young visitors. Activities for kindergartens and primary schools involve educational forms based on *edutainment*, where the children approach the collection through games, touching objects, and drawings. For the students of secondary schools and high school, the knowledge of the collection is deeper and involves the discovery of the object through thematic visits and laboratories where the young visitor may analyze fossils and samples in order to firsthand experience the importance of naturalistic collection for the study of environmental sustainability. Moreover, high-school students may spend a period of internship at the museum in order to understand the operations required for the conservation and enhancement of the heritage and collaborate in the organization of cultural and scientific events.

Another crucial activity for the dissemination of knowledge is the organization of exhibitions and events. During the year, the NHM has hosted inside and outside its spaces, seven large temporary exhibitions covering a period of 302 days, and a total of 35,654 visitors (Table 11.2).

Table 11.2 Large temporary exhibitions hosted in 2018[a]

Temporary Exhibitions	*Location*	*Length (days)*	*Visitors*
Giottolandia	La Specola	58	9,201
Flowers	Botanic Garden	55	5,240
Secret Nature	La Specola	29	3,056
Other Nature	La Specola	53	4,612
Fragments of Nature	La Specola	32	2,565
Flowers of the Earth	External site	9	1,080
Cocoa	External site	66	9,900
Total		**302**	**35,654**

Source: University Museum System, Mission report, 2018.

Notes

[a] Own translations.

Temporary exhibitions have often been focused on understanding how art and nature are connected. 'Giottolandia' has expounded some specimens of calcareous stones with colors and multiform drawings, similar to abstract painting. The stones help to reconstruct the geological history of the earth and to understand its evolution. The exhibition has been accompanied by a series of collateral events that involved educational workshops, a contest, and a conference on the themes of geology. The contemporary art exhibition 'Flower' has been organized in collaboration with the Florence Academy of Fine Arts and has focused on the social topic of geographical borders. 'Secret Nature' has shown to the public a collection of insects' illustrations coming from Malaysia, Indonesia, Australia, New Guinea, and Africa and belonging to the museum's entomological collections. In contrast, 'Other Nature' has been a personal exhibition of an artist working on naturalistic themes and subjects. The 'Fragments of Nature' exhibition has shown the research activity of the museum and university researchers in the Lut Desert, in Iran, in March 2017, for the collection of new meteorites, plants, and animals. The 'Flowers of Earth' exhibition has been realized outside the museum bringing some pieces of the mineralogic collection to a high school in collaboration with the Florentine mineralogical groups and a Geominerary Studies Center. Finally, the 'Cocoa' exhibition organized by the georgophilist academy hosted some specimens from the museum's botanical collection.

Along with these temporary exhibitions, the permanent exhibition 'Mineraliter' located at La Specola has hosted about 18,000 people. The exhibition hosts specimens of the mineralogical collections from Renaissance until today, which have an important scientific value in addition to their aesthetic value. Moreover, a section is dedicated to in-depth studies of the regional mineralogy.

Table 11.3 Events organized by NHM in 2018

Event	*Visitors*
Conferences	850
Meetings	300
Books presentations	250
Theatre performances	1,330
Activities for teachers	100
Other events	900
Total	3,730

Source: University Museum System, Mission report, 2018.

Table 11.3 shows the data for the other events organized by the museum. It counts 121 initiatives implemented on the themes of Natural History, with the participation of 3,730 people.

Finally, a strategy adopted by the museum has been to develop innovative projects with the aim to involve the visitors still poorly engaged in the activities of the NHM. The 'Welcome' project has been dedicated to the disadvantaged, such as people with disabilities, immigrants, or the elderly, to increase cohesion and cultural inclusion. It has been realized in collaboration with eight other local cultural institutions. The aim of the project was to improve accessibility to museums with free guided tours, workshops, and dedicated activities to the disadvantaged. The project 'Whale HUB', on the other hand, has been aimed at including university and high schools students in the activities of the museum. The project has involved the organization of a contest based on the creation of an innovative prototype for audience development in the paleontology section. The goal has been to enlarge the participation of the young public in visiting a new space of the museum dedicated to the marine ecosystem and the theme of environmental sustainability. Three contemporary artists and a research center specialized in studies for the conservation of the marine ecosystem have participated in the project (D'Ambrosio & Dominici, 2019). The project is framed in the wider permanent exhibition "Tales of a whale" – a new expositive space where the NHM hosts a cetacean fossil – that combines the traditional display of fossil of extinct species with the contemporary theme of the protection and sustainable use of the oceans. The final result of "Whale Hub" has been the development of a brochure with augmented reality in order to explore the exhibition and approaching the issue of sustainability with the new technologies. Furthermore, a promotional campaign through social media generated a dialogue platform between museum curators and young visitors (Dominici & Cioppi, 2020).

4.2 Research activities, Publications, and Projects

International research activities have been conducted for the study and documentation of biodiversity in modern ecosystems. These have been developed through field research in countries such as Vietnam, Philippines, Iran, and Albania, in collaboration with other institutions (Table 11.4). Research activities contribute to enlarging the collection of the museum through new samples and to find and describe new species. According to the 2018 mission report data, 57% of new acquisitions derive from the research activity of museum staff, while 43% are donations. Moreover, they are devoted to getting strong collaborations with other international institutions.

According to the mission report of 2018, research activity in Vietnam has been organized in collaboration with the Vietnam National Museum of Nature in Hanoi and allowed the collection of over 1,600 entomological samples and 387 botanical samples in 2018. Within these samples, there are 50 new species. Beyond the acquisition of new samples for incrementing the collection, the aim of the research activity in Vietnam has been also to increase the awareness of the country institutions of the importance of protecting biodiversity, threatened by growing

Table 11.4 The international research activity of the NHM

The focus of the research	*Location*	*Partners*
Conservation of biodiversity	Vietnam	Vietnam National Museum of Nature in Hanoi
Conservation of biodiversity	Philippines	University of Bologna, University of Florence, University of Turin, University of Trieste, University of Almeria, University of Puerto Princesa, Puerto Princesa Underground River National Park
Study of mineralogy and conservation of biodiversity	Iran	University of Florence, Astronomical Observatory of Turin, University of Camerino, Museum of the Sky and Earth of San Giovanni in Persiceto, Shahid Bahonar University of Kerman
Study of botanic and biological classification	Albania	University of Florence, European Union

Source: Authors' elaboration on data from www.msn.unifi.it.

anthropogenic pressure due to rapid population growth, industrialization, deforestation, and pollution.

The research activity in the Philippines has analyzed biodiversity in the UNESCO site of "Puerto Princesa Underground River Cave" and has involved the participation of Italian and international universities. The research activity has been aimed at the collection of samples for the study of biodiversity and also provides a training course in bio-speleology for the naturalistic park staff.

The results of the research activity in Iran have been the collection of 143 meteorite fragments from the desert of Lut in collaboration with the University Shahid Bahonar of Kerman, Iran. The aim of the research has been to study the origin of the Solar System and the biodiversity of the Iran desert in an area still poorly researched.

Finally, through the project 'Agronickel' in Albania, the researchers of the museum have developed an analysis of grasses to study innovative and sustainable processes in biotechnologies for the natural reclamation of polluted nickel sites.

The research activity of the NHM is carried out through the study of its collections. The scientific research conducted by the curators has led to the publication of 86 scientific contributions in 2018. Among these, 30 are scientific articles published in scientific journals with impact factor while 27 have been published in scientific journals without impact factor. Moreover, the researchers of NHM have published 24 books chapters and 5 books, for both scientists and the general public and catalogs. The topics of the scientific contributions have discussed the collections, biodiversity, and specific species but also issues concerning museology, citizen participation in projects on nature, and topics related to pollution or the preservation of ecosystems from the risks of tourism. Finally, the results of scientific research have been discussed at 35 international congresses. A part of the research activity is also carried out through the supervision of degree and doctoral theses.

The collections are visited and studied by the museum staff and by numerous scholars from all over the world. In 2018, 144 external scholars consulted the collections, 58% of the total were academics while 37% came from abroad.

5 Discussion and Conclusions

The main aim of the analysis has been to understand how a naturalistic museum, such as the NHM of Florence, may contribute to creating and disseminating knowledge of science, environment, and biodiversity. The analysis proposed has been devoted to answering the research question "How do naturalistic museums may contribute to creating and disseminating knowledge and making visitors aware of and participate in the issue of sustainability?". It is possible to underline three main results of the analysis.

First, two main categories of activities contribute to the creation and dissemination of knowledge: education and events and research activity. As underlined by Lazzeretti and Capone (2015), studying the event strategy of the NHM of Florence from 2005 to 2011, education and teaching represent a core activity of the museum that spends many efforts in the dissemination of science and biodiversity through visitors' participation. Educational activities, indeed, not only involve the transmission of scientific and naturalistic culture but also promote social inclusion through the removal of physical and cultural barriers. Through its activity, the NHM wants to strengthen scientific competence, promote active citizen participation and action for permanent education in line with the goals of sustainability of Agenda 2030.

In the case of NHM of Florence, exhibitions have both aesthetic and scientific value because they present nature through art and visual representation. Moreover, curators have tried to communicate the value of collection through innovative tools and forms of contemporary art for attracting young visitors. In such a way, museums may become collective educational spaces that convey scientific knowledge, stimulate new approaches, and promote good practices (Hooper-Greenhill, 1999).

Research activities, indeed, mostly concern the study of the collections at museum sites by curators and external scientists, research missions abroad for the study and conservation of biodiversity through the acquisition of new samples, and the communication of research results through the publication of academic articles and books. Research activities on the museums' collections are crucial for their societal value. Science and natural museums, in particular, preserve and enhance biological collections that contribute to public health and safety, monitoring environmental change. Dissemination of knowledge through the study of the collection and the acquisition of new samples are valuable activities of the museums and have an impact on the biological sciences for the study of biodiversity and global climate change (Suarez & Tsutsui, 2004). Moreover, through field research in less developed countries, museums curators may contribute to export knowledge and good practices and make abroad institutions aware of preserving biodiversity as stated by international agreements (United Nations, 2011).

Second, the case of NHM shows that museums interact with different actors and, as a consequence, effective management should consider different kinds of stakeholders. Although the main target of the museum is represented by young visitors, a wider public involves the scientific community, universities, local institutions, and international governmental organizations, such as UE or UNESCO. Long-term relationships between museums and local institutions may have an impact on territories and local development. Museums may become a guide of the environmental and social transformation of the community (OECD, 2017).

Finally, the museum management establishes through its mission to comply with the values concerning sustainable development. In the 2018 social report of the University of Florence, a section dedicated to the University Museum System states that it aims to provide opportunities for reflecting on the interaction between human and nature, giving attention to the cultural formation of the new generations concerning the ecological sustainability. Moreover, in its 2018 mission report, the museum has introduced an explanation of how its activities contribute to the achieving of the goals of sustainable development of the *Agenda 2030 for sustainable development* of the United Nations. This demonstrates that the governance of the museum institutions may promote a long-term program for introducing aspects related to sustainability among the objectives of the institution.

Results of the research evidence that despite the decision to include the themes of sustainable development among the objectives of the museum, concrete efforts to create and disseminate knowledge and to make visitors aware of and participate in the issue of sustainability are still in an early stage. In particular, in relation to the scientific research, the theme is conveyed in a more conscious way and, especially in recent years, the museum curators have made increasing efforts to spreading the environmental issue through scientific and generic publications (Dominici & Cioppi, 2020). However, educational services, thematic exhibitions, and events do not deal directly with the theme of sustainability, but they transmit this message through the intrinsic naturalistic value of the museum's collections. This means that sustainability does not explicitly fall within the themes addressed by the museum in the phase of planification of the activities.

The project that has explicitly dealt with the knowledge dissemination of sustainability was "Whale Hub". Furthermore, if we consider a broader definition of sustainability, which includes the themes of inclusion and social cohesion, the "Welcome" project has shown how synergies with other cultural institutions can contribute to increasing museum accessibility.

Despite these recent efforts, the topic of sustainability has a large consideration within the museum staff. As underlined by the Head of the Botany Section, natural history museums represent "*a focal point to teach environmental sustainability and to activate best practice. They are archives of world biodiversity and preserve the testimony of life on earth in case of loss of biodiversity*". This responsibility is expressed in front of "*school children of all levels. But also, adults in a broad sense*". The recognition of this potential may be a driver for a more comprehensive social change toward the goals of sustainable development.

However, what emerges from the case of NHM of Florence is the difficulty to balance the different dimensions of sustainability. One of the main problems is the old dichotomy between conservation and enhancement. An example is the use of digital technologies. The head of the

Botany Section observes "*digitization may contribute to the dissemination of knowledge; however, it is important to have physical samples*" – and highlights how a low budget may affect the persecution of the museum's mission. Moreover, the head of the Administration Department recognizes that "*there may be a contradiction - between economic, social and environmental value - when there is no funding especially with regard to the physical elements. This is a problem of heritage in Italy in general, therefore funding is necessary to protect and enhance it*".

As stated by the Head of the Zoology Section, the contradictions in terms of generating cultural, economic, and social value can be overcome through a museum's programmatic activity: "*we should understand what to invest in*". This involves the importance of planning the museum operations on the basis of its mission and values. Museums should define specific objectives and timing for operations. In this sense, tools such as accountability and social reporting can also help to monitor the museum's objectives and priorities. As claimed by the Director of the NHM "*the social report can be used to understand the needs and situations of the sites that compose it, but new ideas for enhancement and communication may also arise. The museum has different elements but converging towards the social function*".

The challenge for the future of museums is to attract visitors, increase tourism flows, and maintain financial viability. This should be in balance with the achieving of their institutional purpose (Reussner, 2003). Efforts to overcome this challenge are linked to recognizing the responsibilities to face the risks related to environmental sustainability and to the development of conscious tourism.

Museums are not monoliths but are an expression of the values of society; they reflect its history and customs and how it evolves and adapts to changes (Hooper-Greenhill, 1999). This characteristic makes them crucial places to face the challenges of sustainable development and to educate tourists and visitors about conscious behaviors. From this point of view, management of cultural organizations, even if pressed by the scarcity of resources that requires to implement a strategy based on efficiency and economy, must not forget the fundamental role of society and promote humanistic management oriented to the well-being, cohesion, and social inclusion.

Moreover, the cultural policy may orient funds and investments to sustain those projects that have a value for society and for sustainability. This requires developing instruments to report and account for the impact of museums' activities on sustainable development.

References

Ahern, J. (2011). From fail-safe to safe-to-fail: Sustainability and resilience in the new urban world. *Landscape and Urban Planning*, *100*(4), 341–343.

Alberti, F. G. & Giusti, J. D., (2012). Cultural heritage, tourism and regional competitiveness: The Motor Valley cluster. *City, culture and society*, *3*(4), 261–273.

Andreone, F., Bartolozzi, L., Boano, G., Boero, F., Bologna, M. A., Bon, M., ... & Chiozzi, G., (2014). Italian natural history museums on the verge of collapse?. *ZooKeys*, *8*(456), 139.

Ballantyne, R. & Packer, J., (2005). Promoting environmentally sustainable attitudes and behaviour through free-choice learning experiences: what is the state of the game?. *Environmental Education Research*, *11*(3), 281–295.

Barbagli, F. & Pratesi, G., (2009). *Museo di Storia Naturale dell'Università degli Studi di Firenze: Guida alla visita delle sezioni*. Firenze: Polistampa.

Belfiore, E., (2004). Auditing culture: The subsidised cultural sector in the New Public Management. *International journal of cultural policy*, *10*(2), 183–202.

Bertacchini, E. & Segre, G., (2016). Introduction: Culture, sustainable development and social quality: A paradigm shift in the economic analysis of cultural production and heritage conservation. *City, Culture and Society*, *7*(2), 69–70.

Briassoulis, H. & Van Der Straaten, J., (2013) Tourism and the environment: an overview. In Briassoulis, H. & Van Der Straaten, J. (Eds.). *Tourism and the environment: regional, economic, cultural and policy issues*. Dordrecht: Springer Science & Business Media.

Bulkeley, H. & Betsill, M., (2005). Rethinking sustainable cities: Multilevel governance and the'urban' politics of climate change. *Environmental Politics*, *14*(1), 42–63.

Caloffi, A., Pryke, S., Sedita, S. R. & Siemiatycki, M., (2017). Public–private partnerships and beyond: Potential for innovation and sustainable development. *Environment and Planning C: Politics and Space*, *35*(5), 739–745.

Camagni, R., Capello, R. & Nijkamp, P., (1998). Towards sustainable city policy: an economy-environment technology nexus. *Ecological Economics*, *24*(1), 103–118.

Capone, F. (Ed.). (2016). *Tourist Clusters, Destinations and Competitiveness: Theoretical issues and empirical evidences*. London: Routledge.

Cerquetti, M., (2018). The Importance of Being Earnest. Enhancing the Authentic Experience of Cultural Heritage Through the Experience-Based Approach. In Pencarelli, T. & Forlani, F. (Eds.). *The Experience Logic as a New Perspective for Marketing Management*. Cham: Springer.

Cerquetti, M. & Ferrara, C., (2018). Marketing research for cultural heritage conservation and sustainability: Lessons from the field. *Sustainability*, *10*(3), 774.

D'Ambrosio, V. & Dominici, S., (2019). Whale HUB: Museum Collections and Contemporary Art to Promote Sustainability Among Higher Education Students. In Leal Filho, W. & Bardi, U. (Eds.). *Sustainability on University Campuses: Learning, Skills Building and Best Practices*. Cham: Springer.

Della Lucia, M. & Franch, M., (2017). The effects of local context on World Heritage Site management: The Dolomites Natural World Heritage Site, Italy. *Journal of Sustainable Tourism*, *25*(12), 1756–1775.

Della Lucia, M. & Segre, G., (2017). Intersectoral local development in Italy: the cultural, creative and tourism industries. *International Journal of Culture, Tourism and Hospitality Research*, *11*(3), 450–462.

Dettori, A., Caboni, F. & Giudici, E., (2017). Education and Sustainability: The Case of Emotions Park. *Journal of Management and Sustainability*, *7(4)*, 65.
Di Pietro, L., Guglielmetti Mugion, R., Renzi, M. F. & Toni, M., (2014). An audience-centric approach for museums sustainability. *Sustainability*, 6(9), 5745–5762.
Dominici, S., Cioppi, E., (2020). *Ecce Homo* ovvero i racconti di una balena. In Associazione Nazionale Musei Scientifici. *Passo dopo passo verso la sostenibilità*. Firenze: Angelo Pontecorboli Editore.
Eisenhardt, K. M., (1989). Building theories from case study research. *Academy of Management Review*, *14*(4), 532–550.
Elkington, J., (2013). Enter the triple bottom line. In *The triple bottom line*, 23–38. London: Routledge.
European Council (2014). *Conclusions on cultural heritage as a strategic resource for a sustainable Europe*, Brussels.
Gladwin, T. N., Kennelly, J. J. & Krause, T. S., (1995). Shifting paradigms for sustainable development: Implications for management theory and research. *Academy of Management Review*, *20*(4), 874–907.
Goodland, R., (1995). The concept of environmental sustainability. *Annual Review of Ecology and Systematics*, *26*(1), 1–24.
Greffe, X., (2016). From culture to creativity and the creative economy: A new agenda for cultural economics. *City, Culture and Society*, 7(2), 71–74.
Hooper-Greenhill, E., (1999). *The educational role of the museum*. Psychology Press.
Hunter, C., (1997). Sustainable tourism as an adaptive paradigm. *Annals of tourism research*, *24*(4), 850–867.
Hutter, M. & Throsby, D. (Eds.). (2008). *Beyond price: Value in culture, economics, and the arts*. Cambridge University Press.
Jorgensen, D. L., (2015). Participant Observation. In *Emerging Trends in the Social and Behavioral Sciences*, 1–15. Hoboken, NY: John Wiley & Sons.
Lazzeretti, L., (2012). The resurge of the "societal function of cultural heritage". An introduction. *City, Culture and Society*, 4(3), 229–233.
Lazzeretti, L. & Capone, F., (2015). Museums as Societal Engines for Urban Renewal. The Event Strategy of the Museum of Natural History in Florence. *European Planning Studies*, *23*(8), 1548–1567.
Lazzeretti, L., Sartori, A. & Innocenti, N., (2015). Museums and social media: the case of the Museum of Natural History of Florence. *International Review on Public and Nonprofit Marketing*, *12*(3), 267–283.
Manetti, G. & Valeri, M., (2012). La valutazione del patrimonio museale: il caso del Museo di Storia Naturale di Firenze. *Financial Reporting*, *3*, 111–133.
McKercher, B., (2003). Sustainable tourism development-guiding principles for planning and management. Presentation to the National Seminar on Sustainable tourism Development Bishkek, Kyrgystan, November 5–9.
Melé, D., (2003). The challenge of humanistic management. *Journal of Business Ethics*, *44*(1), 77–88.
Morgan, M., (2004). From production line to drama school: higher education for the future of tourism. *International Journal of Contemporary Hospitality Management*, *16*(2), 91–99.

Mujtaba, T., Lawrence, M., Oliver, M. & Reiss, M. J., (2018). Learning and engagement through natural history museums. *Studies in science education*, *54*(1), 41–67.

OECD (2017). *Culture and local development: maximising the impact.*

OECD (2018). *OECD Tourism Trends and Policies 2018.*

Pencarelli, T., Cerquetti, M. & Splendiani, S., (2016). The sustainable management of museums: an Italian perspective. *Tourism and hospitality management*, 22(1), 29–46.

Pirson, M. A., (2014). Reconnecting management theory and social welfare: A humanistic perspective. In *Academy of Management Proceedings*, *1*, 12245. Briarcliff Manor, NY 10510: Academy of Management.

Reussner, E. M., (2003). Strategic management for visitor-oriented museums: A change of focus. *The International Journal of Cultural Policy*, *9*(1), 95–108.

Sacco, P. L., Blessi, G. T. & Nuccio, M., (2009). Cultural policies and local planning strategies: What is the role of culture in local sustainable development?. *The Journal of Arts Management, Law, and Society*, *39*(1), 45–64.

Sasaki, M., (2010). Urban regeneration through cultural creativity and social inclusion: Rethinking creative city theory through a Japanese case study. *Cities*, *27*, S3–S9.

Scott, C. A., (2009). Exploring the evidence base for museum value. *Museum management and curatorship*, 24(3), 195–212.

Sen, A., (1999). *Development as freedom.* Oxford: Oxford University Press.

Sharpley, R., (2009). *Tourism development and the environment: Beyond sustainability?*. Routledge.

Solima, L., (2017). Museums, accessibility and audience development, In Cerquetti M., *Bridging theories, strategies and practices in valuing cultural heritage*, Edizioni Università di Macerata, 225–240.

Solima, L., (2018). Il gaming per i musei. L'esperienza del Mann. *Economia della Cultura*, *28*(3), 275–290.

Stake, R. E., (2013). *Multiple case study analysis*. Guilford Press.

Suarez, A. V. & Tsutsui, N. D., (2004). The value of museum collections for research and society. *BioScience*, *54*(1), 66–74.

Sutter, G. C. & Worts, D., (2005). Negotiating a sustainable path: museums and societal therapy. In Janes, R. & Conaty, G. (Eds.). *Looking Reality in the Eye: Museums and Social Responsibility*. Calgary, AB: Univesity of Calgary Press.

Thyne, M., (2001). The importance of values research for nonprofit organisations: The motivation-based values of museum visitors. *International Journal of Nonprofit and Voluntary Sector Marketing*, 6(2), 116–130.

United Nations (2011). *Nagoya protocol on access to genetic resources and the fair and equitable sharing of benefits arising from their utilization to the convention on biological diversity*. Montreal: Secretariat of the Convention on Biological Diversity United Nations Environmental Programme. https://www.cbd.int/abs/doc/protocol/nagoya-protocol en.pdf

United Nations (2015). *Transforming our World: The 2030 Agenda for Sustainable Development.* https://www.un.org/ga/search/view_doc.asp?symbol=A/RES/70/1&Lang=E

UNWTO (2019). *World Tourism Barometer and Statistical Annex, 17(1), 1-14.* https://www.ttr.tirol/sites/default/files/2019-02/UNWTO%20Barometer%20Vol.%2017%20%28J%C3%A4nner%202019%29.pdf

WCED (1987). *Our Common Future (Brundtland-Report).* New York/Oxford: Oxford University Press.

WTTC (2019). *Travel & tourism Economic impact 2019.* London: WORLD TRAVEL & TOURISM COUNCIL (WTTC).

Yin, R. K., (2017). *Case study research and applications: Design and methods.* Sage publications.

12 How Museums can Contribute to Collective Well-being

The Role of an "Augmented" Video Guide

Adele Magnelli

1 Introduction

Daniel Bell, a 20th-century American sociologist, stated that "technology, like art, is a soaring exercise of the human imagination" (Bell, 1980, p. 20). This statement points out how technology promotes imagination and investigation, generating new ways of thinking about the world. The term technology comes from the Greek *tékne* "art" and, according to Bell, the two fields should be considered together. In this sense, technology is "a form of art that builds a bridge between culture and the social structure while remodeling both in the process" (Bell, 1976).

Art and technology are, therefore, categorized as two dimensions connected to each other by a common thread and giving life to a single essence aimed at improving the human processes of idea creation and widespread knowledge. Imagination has a key role in generating new, dynamic, and constantly evolving creativity. This process also includes the concept that culture is no longer reserved for a select few, but, thanks to the diffusion of technology, it has become accessible to a much wider audience, facilitating communication, increasing the level of engagement, and creating memorable experiences. It can also contribute to the level of user psycho-physical well-being, in terms of well-being as an output of fulfilled self-esteem needs.

In the current age, digital technology has completely changed the way in which we interact with museum institutions. Compared to the past, users have become increasingly inquisitive, hungry for knowledge, and thanks to the web and to participate in forums, blogs, and social networks, they are well-informed and able to recognize the truthfulness of the information circulating on the web.

Consequently, before going to visit a museum, people very often inquire about what they will see, developing higher expectations for broader and more complex proposals and driving cultural-sector operators to come up with ever more attractive and high-value solutions. The next step in this logical process is the development of a paradigm of innovative museum visit/experience, based on the evolution of the tools that the museum itself has at its disposal. The aim should be enhancing the ability of museums' sites to offer an enjoyable, interesting, and educational experience to the curious visitor. Maslow theorized that a human being's needs are hierarchically structured in a pyramid (Maslow, 1943); at its base, lie the psychological needs, upon which all others stand further up until we arrive at the pinnacle of self-actualization. The basic importance of psychological need as a basis for human happiness and well-being underlines the extreme importance of socially enriching experiences.

Positive psychology is concerned with *eudaimonia*, "the good life", reflection about what holds the greatest value in life – the factors that contribute the most to a well-lived and fulfilling life. While not attempting to provide a strict definition of the good life, positive psychologists agree that one must live a happy, engaged, and meaningful life in order to experience *the good life* as defined by Martin Seligman as "using your signature strengths every day to produce authentic happiness and abundant gratification" (2019, "Authentic Happiness", p. 20).

The concept was also developed in Flourish: A Visionary New Understanding of Happiness and Well-being (2011), where Seligman approached the idea of "meaningful life" as depending on five different but complementary elements. These, strictly linked with the five elements of his well-being theory, form the acronym PERMA:

- **Positive** emotions, frequently connected to positive outcomes in life;
- **Engagement**, as passion and concentration in dealing with the task at hand. Mihaly Csikszentmihalyi (1990) explains true engagement as a flow, a feeling of intensity that leads to a sense of ecstasy and clarity.
- Relationships, both inside and outside family, both physical and platonic, which fuel important positive emotions;
- Meaning, which helps in putting everything into context and to compare one's existence with the bigger picture;
- Accomplishments, to be pursued even when not resulting in a positive outcome. However, it must be noted that they can also activate other elements of PERMA.

Well-being, according to Seligman (2011), thus, depends on a life that is full of emotions and feelings, engaging, exciting curiosity, and structured after a personal path of growth; personal involvement and immersion are powerful tools for a museum that seeks to play an important part for human well-being. More specifically, cultural heritage museum experiences can be designed to convey positive emotions by providing paths that involve not only information and notions but – most importantly – a first-person point-of-view of cultural themes such as the artist's life and feelings. VR has the ability to induce a feeling of "presence" in the computer-generated world experienced by the user (Riva et al. 2007).

The approach of Humanistic Management (HM), rooted in the great humanistic and Renaissance heritage of Italian and European culture, is part of this discourse. It uses the tool represented by the opening not only toward areas that the 'scientific' enterprise has always considered alien to itself, such as the different forms of art, but also toward new territories unlocked by new digital and multimedia technologies. Those pertaining to the New Economy, understood not only as a mere aggregate of companies that have the internet and Information and Communications Technology (ICT) as their main business and not even as all companies that provide new technologies, such as ETT, but also as a set of companies that those technological innovations somehow use, the user sectors, including museums, containers of excellence of the world's artistic heritage. In this regard, since for HM, art is conceived as an operational tool aimed at overcoming sectorial languages and is a multidisciplinary catalyst that has the task of spreading knowledge, new technologies, such as Augmented Reality in the museum field, at the center of the speculation of this chapter, can play a very important role.

In terms of "engagement", multimedia technologies represent a useful tool to renew visitors' curiosity as a purpose for cultural experiences. Storytelling, a discipline that uses the principles of rhetoric and narratology in the process of organization and presentation of content also through digital technologies, is an essential part of this innovative paradigm: by carefully preparing not just interesting contents, but also an involving and enticing way to propose them, we provide a powerfully communicative content that can thus reach more visitors through technologically adept means.

For some months now, in this perspective of renewal, the Musei Civici di Roma Capitale (the Capitoline Museums, the Ara Pacis Museum, and Trajan's Market-Imperial Forum Museum) have been equipped with a highly innovative video guide. It was created by ETT S.p.A., that won a tender for Zètema Progetto Cultura, in collaboration with the Capitolina Superintendency for Cultural Heritage.

This multimedia tool combines the characteristics of a traditional audio-guide with the use of augmented reality (AR), which enriches reality with a whole series of information to superimpose on what the eyes see, and virtual reality, and (VR) a realistic simulation of a reality that does not exist. Museum exhibits are described by means of intelligent storytelling.

The video guide can be described as a high-potential visit tool, able to diversify the user experience, sharpen curiosity, and encourage further learning. The application may be used either on video guides available at the three museums or downloaded onto personal mobile devices together with content that may be seen at home. This is the real new feature of this innovative tool, as users can continue to enjoy museum content after the visit, thereby increasing the pre- and post-visit experience. This type of custom-designed tool contributes to a higher level of user well-being and involvement.

In the three museums, visitor localization is guaranteed by a system based on BLE Beacon technology, small Bluetooth signal-repeaters installed near points of interest (POIs). Content relative to a nearby POI is automatically suggested, thus reversing traditional interaction where the user must type in a number on the audio-guide or frame a QR code to access in-depth content.

This way visitors, even those unaccustomed to technology, can easily enjoy the content and find their way around a museum. Thanks to this sustainable technology, there is no physical impact on museum spaces. Moreover, the video guide lets visitors choose among various predefined thematic routes, guided by engaging storytelling that fully involves them. Visitors, therefore, become active users, no longer passively looking at the content on offer, and are able to enjoy a highly formative and stimulating experience.

In this chapter, starting from a detour to look at museums today and on the role of technology, the evolution of previous-generation audio-guides into modern video guides will be analyzed. In this paper's case study, we shall show the "Musei in Comune" video guide, followed by a section on the innovative aspects of this instrument and how they have modified a museum visit experience.

New media is a continuously evolving trend and digital ecosystems offer opportunities to enrich a visitor's experience. All this is closely connected to other new museum trends, such as inclusion, intercultural dialogue, social commitment, creativity, and participation.

2 Museums Today and the Role of Technology

The cultural sector has gone through an intense growth in recent years. This leads to new considerations relating to the different and new ways of showing and narrating objects and past events, interconnected with suggestions developed from studies on culture, tourism, and ICT.

Today's museums and cultural-sector companies are at the center of a thought process that is revising its criteria on the ways that collections and past events are organized, made accessible, and communicated. At the same time, they are reflecting on the ways in which modern museums can be interpreted. They are no longer mere containers that simply store, preserve, and display objects; they are spaces for democratic and inclusive growth and confrontation, serving various categories of users who have new needs and new expectations (Bonacini, 2011).

In modern multicultural societies, museums need to renew their relationship with visitors and users, becoming cultural expansion centers that develop and extensively disseminate culture. Visitors, therefore, have a certain intrinsic dignity; a result of simply being human but also acquired through action, such as – for example – during a museum visit.

As Scartabelli (2014) maintains, "a museum today is no longer a simple place for exhibiting works and a container for objects, but in addition to its conservation role, a museum also has a social role. Today's museums aim to become real hubs of activity, culture, research, and study. They intend to be of fundamental importance to modern society, having a privileged relationship with schools and museum didactics by skillful use of what new technologies can offer" (p. 45).

To summarize, museums have gradually grown from the simple role of preserving and giving information to that of places for learning through experience. Thanks to new technologies, this process is increasingly emotional.

A museum, by virtue of its strong social vocation, can contribute to the spread and improvement of the well-being and strategies increasingly present in our lives. Knowing how to talk, how to communicate, and how to transmit clear, certain, and verified information means giving each cultural visitor the right to an enjoyable path through places of culture so that everyone can find "their" museum.

It is true that today a museum is an active and proactive cultural institution with the task of encouraging contact between the public and the materials it holds, stimulating a revival of the memory of the past, and leading to dialogue with modern reality (Scartabelli, 2014). This way, a museum is a modern institution, open to innovation, and able to actively contribute to the development of thought and culture in contemporary society. Its basic functions remain unchanged – protection, conservation, enhancement, and promotion – but these are now profoundly modified and assisted by technological progress and the consequent introduction of new tools that guarantee the enormous potential for the development of the whole sector.

These newly rethought museum spaces, thanks to specially designed multimedia solutions, adapt to particular surroundings. They more often than not use soft technologies, invisibly welcoming and accompanying users during the visit experience. In this way, visitors can focus more on interacting with content rather than on technological tools. This, for example, is the case when augmented reality is used in museums to give additional information levels on collection pieces or to display digital versions of artists next to their works. Augmented reality means that a third dimension and digital content (data, audios, videos) may be added to displays, giving life to objects and events. As Coates writes in a 2019 article published on the portal of the MuseumNext international conference series, many museums around the world already use augmented reality to give a new point of view to existing collections, thus attracting a wider audience, like the Smithsonian Institution in Washington DC. In

2017, the Smithsonian created an app called Skin and Bone with which thirteen animal skeletons could be reconstructed. Visitors see the animals moving, giving them a new way of viewing the museum collection. They can use the app to see, for example, a bat take flight or a large Anhinga bird feeding.

Another good project example was created at The Art Gallery of Ontario in Toronto. In 2017, the museum collaborated with digital artist Alex Mayhew to create an augmented reality installation called ReBlink. Mayhew reinvented some of the existing pieces in the collection, in which the subjects of paintings come to life and are transported into 21st-century reality (Coates, 2019).

These examples show how an effective adaptation of museum presentation to the varying cognitive expectations of users imposes an adaption of both the museological project and the museum visit route. The collections on display need to be integrated with all the associated information necessary for the enrichment of the visitor experience. To achieve this goal, museum visit routes are increasingly enhanced by the coordinated presence of audio-visual supports, animations, simulations, and sound accompaniment. These are used both to enhance the exhibition setting and to present additional information or in-depth content (Bonacasa, 2011).

Cultural heritage has a great need to communicate, to explain. It contains a universe of stories and parallel emotions that intertwine with each other and with the works on show. It is evident, therefore, that the main duty of a museum is to make the most substantial and underlying meanings of the works known to visitors. The technical task of sector operators will be to minimize the distance between exhibits and visitors, using communication tools to help them better understand these meanings and stimulate them to think about and to desire knowledge. In the past, explanatory texts, guides, catalogues, captions, panels and guided tours performed this function, and in part they still do, but the limits of these tools are obvious, as they are unable to explore such a broad universe of meaning. In this sense, new technologies can certainly help because these days communicating does not mean just giving knowledge using panels or captions, but providing an authentic experience, making a museum a place to be known, understood and lived, in which the visitor becomes the central character of the visit and the discovery experience. First person experience, using all five senses, is the best way to remember a series of data, concepts and information. These will not be forgotten, rather, emotion will make them akin to a precious treasure chest of knowledge to dip into.

Furthermore, making cultural places and heritage accessible makes them useful to improve people's lives. Social welfare (Social Welfare Function, SWF) has become one of the indexes analyzed by modern economics, which examines "that dimension of well-being of an individual that concerns his relationships with others, how other people react to him and how he interacts with social institutions and the norms of society" (Mandosi, 2017). Social well-being highlights another important issue relating to the role of museums: ethics. The International Council of Museums (ICOM) already addressed this issue in 1986 and produced a Code of Ethics (revised in 2004). This established the importance of museums as places having strong responsibilities towards society. Much like access, clarification and education, they must aim to actively involve communities to whom the preserved heritage belongs, interacting with them in the phases of protection and enhancement (Mandosi, 2017). Nowadays a museum is an open environment ideal for meeting people, communicating with the local territory, stimulating local and global knowledge and economies. It plays a decisive role in responding to identity crises, creating self-aware and responsible people. A museum is open to all those who wish to get involved, face modern challenges and work on building tomorrow, without forgetting the past.

3 The Experience and Audience Development

Since the 1960s, "experience", as a concept, has increasingly received worldwide academic recognition and attention, and it has been given many intrinsic definitions and meanings.

In the cultural sector, users now demand more exciting and immersive experiences, not only when purchasing products and services, but also when they are being used and enjoyed (Pine and Gilmore, 1999).

According to these two scholars, an experience is a new type of offer that can be added to commodities (fungible materials), goods (tangible products) and services (intangible assets). Experiences are unforgettable events that personally involve the consumer at that moment. They may be classified depending on active or passive participation and how people connect with the event.

Pine and Gilmore (1999) speculate that there are four types of experience:

1 Entertainment experience: the person uses their senses to passively take in what is happening;
2 Educational experience: the person takes in the experience with active body or mind participation;
3 Aesthetic experience: the person is physically immersed in an experience but remains passive;
4 Escape experience: the person is totally immersed in the experience, with active participation.

Schmitt (2006), on the other hand, sees the classification of experiences through five Strategic Experiential Modules (SEM):

1 Sense: the experience involves the senses;
2 Feel: the experience involves feelings and emotions;
3 Think: a cognitive experience;
4 Act: physical experiences;
5 Relate: experiences that depend on interactions and relationships with others.

Experiential marketing is therefore conceived as a process involving all of these elements. It is necessary to attract attention first through sensory experiences, then to stimulate or recreate specific moods and feelings associated with the product, create stimuli connected to people's intellect and problem-solving abilities, invite action, and put the person into perspective with himself and with other cultures (Schmitt, 2006).

As a result, during the last few decades, experiential marketing has also influenced cultural sector institutions in their offering of unique and memorable experiences.

An experience, in a cultural setting, is a dynamic process that evolves over time and space. It can begin before the museum visit and continue well beyond it, as it has been enriched with sensorial, emotional, cognitive, behavioral, and relational values, as well as dimensions linked to social sharing and the value given to the transmission of information.

Another challenge for the cultural sector is linked to the concept of audience development, conceived both in a qualitative sense, to strengthen existing cultural demand, and in a quantitative sense, involving new and potential audiences with the view to expanding the social base of cultural enjoyment.

This objective, outlined in the Creative Europe framework program, refers to the fragmentation of European cultural markets and the ever-increasing need to expand the audience for cultural products, also through experiments with new forms of involvement achieved by using digital media.

Cultural participation is becoming progressively more hybrid and multidimensional. It is clear that an audience-centered approach is now fundamental if cultural institutions are to realign with their audiences and overcome the traditional top-down logic for knowledge transfer by creating cultural value with the public. Analysis of usage data and online feedback can influence the evolution of exhibition processes.

4 The Evolution of Previous-Generation Audio-Guides into Video Guides

One of the museum tools used to help a visit, the audio-guide, has been progressively renewed to offer highly qualifying experiences. Introduced with analog technologies during the 1960s and subsequently updated in the 1990s to a digital format, it is among the most widespread and simple instruments in museums, cultural places, and archaeological sites. These are generally earphone devices, for individual or group use, on which visitors hear a pre-recorded voice providing information on the exhibited objects and works, without the distraction of having to read information panels. Various types of audio-guides exist, the most widespread and simple being basic multimedia mp3 file players. The internal memory of the device has a voice recording accompanying visitors on a predetermined route. This type of device is generally provided by the museum, for free or on a lease, and offers some visit-customization possibilities, such as the choice of language, route, and duration of the visit (Scartabelli, 2014).

Audio-guides have the advantage of being able to store vast quantities of text that would be difficult to show on panels or in captions. They are widely used in museums and temporary exhibitions throughout the country and very few museum don't use them. They are particularly popular with foreign tourists, who often find this system to be one of the rare interfaces designed to communicate in their native language.

Along with the classic audio-guide, developments in technology have led to the spread of increasingly innovative devices, such as video guides, tools that present visual content such as images and videos, combine augmented reality and virtual reality experiences with audio content on various devices, such as smartphones and tablets, like the one in this chapter.

Technological development of these tools is not the only evolutionary factor. Museums are now committed to structural enhancement in order to increase the level of engagement and develop communication methods, as well as the relaunch of their cultural and social functions designed to improve user well-being. This commitment does not just improve the display of museum collection exhibits but also tends to give a better understanding of the collections themselves. This changes the level of information offered to visitors and consequently leads to the creation of a different and substantially innovative visit experience (Bonacasa, 2011).

This is why the video guide created for the Musei Civici di Roma should be looked at as an innovative tool that is capable of breaking down possible barriers between visitors and the cultural site. In this context, the technological contribution simply adds value to the museum visit, enriching it with new interactive and multimedia experiences.

5 The Video Guide: Museums, Storytelling, and Visit Route

The following use case revolves around the new video guides developed for three museums in Rome: the Capitoline Museum, the Trajan Market Imperial Forum Museum, and the Ara Pacis Museum, all part of the

Rome Capital museum system, comprising extremely different sets, museums, and archaeological sites of immense cultural value.

The Capitoline Museums opened in 1734 under Pope Clement XII and are considered to be the world's oldest public museum complex. They constitute the most important civic museum organization in Rome. The oldest part of the collection dates back to 1471, when Pope Sixtus IV donated to the Roman people the bronze statues of the Lupa, the Spinario, the Camillo, and the colossal head of Constantine, with the globe and the hand. Initially placed on the external façade of the Palazzo dei Conservatori courtyard, the statues were joined by further archaeological finds, closely connected with the history of ancient Rome. They are now exhibited in Palazzo dei Conservatori, Palazzo Nuovo, and the Pinacoteca Capitolina.

A unique archaeological complex in the world, Trajan's Market went through the evolution of the city since the imperial age and was constantly reused and transformed; first as the strategic administrative center of the Imperial Forums, then a noble residence and military fortress during the Middle Ages, to become a prestigious convent during the Renaissance and afterward barracks after the fall of Rome. Today, the area of the Markets hosts the Imperial Forum Museum, dedicated to the architecture and sculpture of the Imperial Forums. Today, the museum exhibits architectural reconstructions of the Forum of Caesar, the Forum of Augustus, the Forum of Nerva, Templum Pacis and Trajan's Forum.

The Ara Pacis Museum, designed by Richard Meier and inaugurated in 2006, houses the Ara Pacis Augustae, the altar dedicated in 9 BC by Emperor Augustus to the Pax Romana, the divinity symbolizing the peace and prosperity of the Augustan era. When first rediscovered, it was inaugurated by Mussolini, in 1937, in the Morpurgo display hall between Via di Ripetta and Lungotevere, where the new museum is.

> "You know what? This is the world's first public collection. It was Pope Sixtus IV who, in 1471, transferred four famous bronze statues from the Lateran Palace to this building, the Palazzo dei Conservatori, built some decades earlier for the city magistrates. The sculptures are the famous Lupa, the Spinario, the Camillo and the large head of Constantine with the hand and the globe; works that you can admire during your visit".
>
> (Neutral narrator, Capitoline Museums.)

> "Hello, my name is Apollodorus and I'm the Empire's most famous architect. Emperor Trajan commissioned me to carry out all kinds of projects: bridges, war machines, triumphal arches, ports, spas and, of course, the forum that will pass on his name to future generations!"
>
> (Apollodorus, Trajan's Market.)

> "...Oh, I'm so sorry, I haven't introduced myself! I'm Pax. Yes, Pax, the goddess of peace. It is to me that Emperor Augustus dedicated the altar that you see at the end of this enormous room of glass and metal."
>
> (Pax, Ara Pacis Museum.)

These are a few examples of the approach that was chosen to immerse the visitors in the atmosphere of the museums. The virtual guides that accompanied them through history and knowledge are historical figures who lived through the events that they are showing to the visitor, now a traveler through time and space; the events that contributed to the foundation of the very museum they are in propagate like an echo, throughout the marbles, the works of art, to reappear in front of the user.

In particular, while the wide amount and type of contents of the Capitoline meant that the male and female narrators had to be generic, while still powerfully communicative, the other two museums' highly focused narrative allowed for exceptional narrators. Trajan's Market houses the digital phantoms of the emperor Trajan and the architect Apollodorus, embodiments of different points of view and owners of deep knowledge of every column, capital and statue.

On the other hand, the Ara Pacis Museum's visitor will be greeted by none other than the titular host of the temple, the god Pax, made official by Augustus with the building and presentation of the Ara, embodiment, and witness to the long era of peace and prosperity envisioned by the emperor, master of every secret hidden in the nooks, crannies, decorations of the monumental temple.

These choices were not based on style exclusively; rather, the theory behind the concept of immersive storytelling demanded a powerful, emotional figure to lead them and involve the user, while at the same time it had to be someone who had every reason to possess complete knowledge of the history of the place.

This calls for even more attention to the choice of the narrative style; it must explain clearly a number of complex and often abstracts concepts, while maintaining involvement of the user and all this while maintaining the characterization of the narrator's persona.

This also meant underlining the peculiarities of the different museum's personality and character: the Capitoline Museums, an exciting stage rich in history both in its structure and as a host to one of the oldest collections in Rome; Trajan's Market, the scene of numerous transformations and historical stratifications that have resisted the course of time for over two thousand years and incorporate the most important remains of the Imperial Forums; The Ara Pacis Museum, the main symbol of Augustan politics and which is revived today, thanks to Richard Meier's new museum arrangement. The video guide routes have been designed and structured according to the historical and archaeological characteristics of each museum. They represent technology's answer to the various requirements of increasingly diverse cultural audiences.

This is proof of the importance of storytelling: choosing how the cultural concepts and notions are offered to the visitor is crucial to respect both the character of the location, the need for historical and cultural clarity and precision, and the need for a more entertaining visit for the user.

Visitors may choose a route among those on offer, depending on how much time they have and what they would like to learn more about. A complete itinerary has been designed for each museum that includes all the works and is the subject of an in-depth study by the video guide. Other routes go into more detail on specific themes, such as the ancient world's main characters and the human and divine events of legend and myth, the architecture of the Imperial Forums, or the hidden meanings of the Ara Pacis reliefs.

The video guide also has a section showing current exhibitions on the City Museums circuit (MIC). It is constantly updated in order to increase public involvement of what's on locally so that people can see cultural realities that might otherwise be missed.

6 The Innovative Aspects of the Video Guides: Beacons, Augmented Reality, and Virtual Reality

Another technological advancement that is crucial to immerse the video-guide user, specifically by minimizing the need of physical input on the device on his part, is the beacon, a physical object placed on specific parts of the exhibition and used to automatically present Points of Interest (POIs). These shall be selected among those that hold particular historical and cultural significance, so that the user will be guided toward them by his natural curiosity and the personal involvement stimulated by storytelling.

The beacons are low-frequency Bluetooth signal repeaters (BLE transmitters) using a button-type battery, with a transmission range from 1.50 m to 70 m. They are practically unnoticeable and easy to install and have completely revolutionized the way in which we interact with museum spaces by allowing for specifically localized multimedia content on museum works or a monument. Indeed, the video guide itself will interact with the beacon; when close enough, the beacon's signal will unlock specific content on the device.

Content is presented to visitors with almost natural interaction. Even people not accustomed to using technology can easily enjoy the multidisciplinary in-depth analysis. They can independently decide the level of detail for their personal discovery experience, and they don't need to keep trying to get their bearings – even in a large museum.

A video guide shows innovative augmented reality content; this means that the users will not isolate themselves in a virtual world. Rather, they will experience the real world around them, while the videoguide will provide additional multimedia content that is organically and strategically "placed" on the individual POIs in a new and immediate way. Focusing the device camera on the exhibit can indeed trigger additional information levels, such as in-depth videos, historical photos, reconstructions, or text information.

Imagine that you are entering the Capitoline Museums, crossing the courtyard in the Palazzo dei Conservatori where you see the surviving pieces of the colossal statue of Constantine, then climbing the monumental staircase before pausing in the Sala dei Capitani (Hall of the Captains). Here you see a triumph of frescoes; centuries of talking history from the early period of the republic, in which examples of military value and civil virtues are narrated, perfectly continuing the story of the Sala degli Orazi and Curiazi. While in this room, point your device camera at the *Frescoes by Tommaso Laureti*. You will learn the names of the characters depicted on the frescoes. Further on, you pass the famous Sala della Lupa (Hall of the She-wolf), which holds, like a priceless treasure chest, the Capitoline Wolf – the symbol of Rome. You will then stop in front of a precious ceremonial wagon, the *Tensa Capitolina*, made from bronze plates and rich in friezes depicting scenes from the life of Achilles until the siege of Troy. From a static image of the Tensa, we move on to a moving one showing the most emblematic episodes represented on the cart. In the Picture gallery inside Palazzo dei Conservatori, you can admire paintings of exceptional importance, such as *Romulus and Remus* by Peter Paul Rubens and *The Fortune Teller* by Caravaggio. For the first one, augmented reality follows the detailed audio track and you progressively learn about the characters of the painting that depicts the myth of the birth of Rome, with the she-wolf suckling the twins, Romulus and Remus. In the second case, by rotating the tablet horizontally, you will discover a radiograph of the preparatory sketch concealed behind Caravaggio's painting, showing how the orientation and subject were different from what can be seen today. From Palazzo dei Conservatori, you go toward Palazzo Nuovo. Here you will find the *Galata*, the proud but prostrate warrior, commissioned by Attalus I of Pergamum to celebrate his victory against the Galatians in the third century A.D. By aiming the camera at the AR logo, positioned next to the statue, the augmented reality experience will show you some details of the statue and the various copies of the Galata existing in different collections.

After entering the Trajan's Market-Imperial Forum Museum, the Great Hall is to be found. It is one of the buildings of the monument and here, thanks to augmented reality, a reconstruction dating back to Roman times may be seen. Walking through this space, you reach the Introductory Room of the Forum of Caesar, where a Frieze with figurines of cupids is kept. It was part of the interior decoration of the

Temple of Venus Genetrix. By aiming the device camera, the relocation of the fragment in its original context may be seen. After going up the stairs to visit the first floor, with the remains from the Forum of Caesar and the Forum of Augustus, you should go to the Belvedere Terrace where, in addition to enjoying an exceptional view of the Imperial Forums, a series of insights on the Forum of Caesar, Trajan's Column and Basilica Ulpia may be seen. After seeing the terrace and walking along Via Biberatica, an ancient Roman road that still survives, you can go down to Trajan's Forum, enter the northern Frontal Hall and approach a frieze with sphinxes. Thanks to augmented reality, you can admire the reconstruction of the whole figure.

The journey is not yet over; the last stop is the Ara Pacis Museum. Inside the entrance, there is a model of the Campo Marzio. Aiming the device camera activates a series of in-depth features on the main details present in the area at the time of Augustus: the Ara Pacis, the Saepta Iulia, the Mausoleum of Augustus, and the Horologium. Continuing the visit, a series of busts belonging to the members of the Julio-Claudian family may be seen before stopping in front of the Bust of Augustus. Here, by aiming at the AR logo, a video shows the key stages of his life and his ascent to the throne. The Ara Pacis, that symbol of the Augustan era, can then be seen in all its magnificence. By going around the Ara and stopping behind it, you can point the device camera at the right-hand floral frieze and admire the original colors of the frieze as they would have appeared in the Augustan age.

Recapping, there are four different modalities through which we are currently exploiting the potential of AR to empower the visitor's experience through our AR video guide: *discovery mode,* through explorable, in-depth features that give text information and facts on details of the works; *narrative/animation/ AR planar mode,* for some paintings, the audio narrative follows the AR animation that reveals details of the work; *pure recognition*, recognition of the work activates related content; *AR logo,* by aiming at the easily recognizable AR logo, visitors can activate videos relating to some of the museum collection exhibits and get additional information and content.

The three museums may also be explored through exclusive 360-degree virtual reality content, accessible from the "panoramic" section on the apps that can be downloaded from app-stores and seen through Cardboard viewers. The content includes panoramic photos, new and suggestive perspectives providing a new way of exploring and promoting museum spaces. The use of such content can give a preview of the visit, creating desire and curiosity, but it can also be used after the visit, acting as a novel and innovative souvenir.

To ensure accessibility for users with hearing disabilities, the video guide also shows adapted and simplified text for deaf people, thereby lowering social barriers.

7 Conclusion and Possible Future Scenarios

The constant evolution of a museum's living tissue, so to say, was faced with the challenge of adapting a fixed and rigid structure, which entailed great costs and logistic issues. This issue has been beaten by this technologically advanced approach. Today, whoever is better at capturing the attention of the public wins, putting people at the center of a museum visit experience and helping them raise their state of well-being and engagement. This was the driving force behind this project; using a custom-designed tool to amplify the feeling of the psychophysical well-being of visitors, making them feel fully involved.

Consequently, the video guide that we have produced seems to be in line with the function and social role of a museum, seen as a place with strong responsibilities toward society, including those regarding access and interpretation as well as education, with the aim of actively involving communities to whom the protected heritage belongs. Such involvement was made possible thanks to a multi-faceted tool like the device described, which has revolutionized museum visit tools.

It is one thing to tell a story, but entirely another to be able to see it. This device leads visitors to the discovery of history, through their own senses and feeling of wonder.

In this perspective, we wanted to put the person, with their social values and cultural background, at the center of the experience. "People ignore design that ignores people" (Frank Chimero, Designer). This statement reaffirms the importance of Experience Design and putting the user first. Any aspect of an experience that does not consider user needs is bound to fail.

In addition to contributing to a state of well-being in visitors, the video guide embodies innovative possibilities, such as the use of a Beacon BLE system, overturning traditional interaction with technological tools, and combining different levels of in-depth customization with indoor localization. In museums, visitors no longer have to type in numbers on traditional audio-guides or search for QR codes. Now, it is the content that follows the visitors, locating them in museum rooms and, almost as if by magic, appearing on the video guide, making the museum visit much easier. The simplicity of the use of the video guide means that visitors who are following one of the proposed routes can move freely around the museum, feeling free and not confined by technology. Storytelling involves visitors in what is going on, allowing them to identify with the story and memorize abstract and complex concepts. Emotional involvement is boosted thanks to the astute choice of using, as storytellers, characters who really existed, who express their enthusiasm, curiosity, and points of view on the story as it unfolds. Various modes of communication range from first-person accounts to the use of dialogue and museum-theatre techniques. The use of augmented reality (AR) further enriches the visitor experience at the three museums, astounding them. So, for example, from a static image of the Tensa Capitolina, we can go to a moving one in which a preparatory sketch will emerge from a Caravaggio painting, giving it a new interpretation, and then to ruins of the Great Hall of Trajan's Market, where visitors can see the superimposed original structure.

The main expectation for this tool was the ability to go beyond the classical concept of just telling a story, and give a 360-degree view of history, with visitors being active users.

We are now at the point of wondering what could be possible future scenarios, given that the evolution of ICT goes hand in hand with historical and archaeological research. Future hardware architecture, designed for the three museums, could make prosumers more active by taking part in the production of museum processes and, for example, co-creating content.

A further development could be linked to emotion analysis. This would register the emotions and feelings of visitors while they are looking at a particular work of art, or be based on the average time spent in a room. Depending on the data received, visitors could be profiled according to their preferences and would receive notifications with messages, thus making a museum visit even more participatory by adding a combination of surprise, experience, and emotion.

We can also think about possible future visitor interaction with Artificial Intelligence systems that can learn and evolve in order to communicate art and culture.

References

Addis, M. (2002). *Nuove tecnologie e consumo di prodotti artistici e culturali: verso l'edutainment. Micro & Macro Marketing*, XI(1), 33–59. doi: 10.1431/2991:y:2002:i:1:p:33-60

Antinucci, F. (2004). *Comunicare nel museo*. Bari, IT: Laterza.

Balboni Brizza, M. T. (2007). *Immaginare il museo. Riflessioni sulla didattica e il pubblico*. Milan, IT: Jaca Book.

Bell, D. (1976). *The Coming of Post-Industrial Society: A Venture in Social Forecasting*. London, UK: Penguin, Harmondsworth.

Bell, D. (1980). *The Winding Passage: Essays and Sociological Journeys 1960-1980*. Cambridge: ABT.

Bonacasa, N. (2011). *Il museo on-line. Nuove prospettive per la museologia*. Palermo, IT: Osservatorio per le Arti Decorative in Italia "Maria Accascina".

Bonacini, E. (2011). *Nuove tecnologie per la fruizione e valorizzazione del patrimonio culturale*. Rome, IT: Aracne.

Camarero, C., Garrido, M. J. & Vicente, E. (2015), Achieving effective visitor orientation in European museums. Innovation versus custodial. *Journal of Cultural Heritage*, *16*(2), 228–235.

Cataldo, L. (2011). *Dal Museum Theatre al Digital Storytelling. Nuove forme della comunicazione museale fra teatro, multimedialità e narrazione*. Milan, IT: Franco Angeli.

Coates, C. (2019). *How Museums are using Augmented Reality*, Retrieved from https://www.museumnext.com/2019/02/how-museums-are-using-augmented-reality/

Csikszentmihalyi, M. (1990). *Flow. The Psychology of Optimal Experience*. New York, NY: Harper & Row.

Fisher, J. (2004). Speeches of Display: Museum Audio Guides by Artists. In Drobnick J. (Ed.), *Aural Cultures* (pp. 48–61). Toronto, CA: Toronto & Banff: YYZ Books & Walter Philipps Gallery Editions.

Gretzel, U. & Tazim, J. (2009). *Conceptualizing the creative tourist class: Technology, mobility, and tourism experiences. Tourism Analysis. 14*(4), 417–481. doi: https://doi.org/10.3727/108354209X12596287114219

Guccione, M. (2012). *Come sarà il Museo del futuro?* Rome, IT: Maxxi.

Hodson, R. (2001). *Dignity at work*. Cambridge, UK: Cambridge University Press. doi: https://doi.org/10.1017/CBO9780511499333

Jalla, D. (2000). *Il museo contemporaneo. Introduzione al nuovo sistema museale italiano*. Torino, IT: UTET Università.

Lazzaro, A. (2017). *Innovazione tecnologica e patrimonio culturale tra diffusione della cultura e regolamentazione. Atti del Convegno di Messina 14–15 ottobre 2016, in* www.giustamm.it., 2017

Lazzeretti, L. & Sartori, A. (2016). Digitalization of cultural heritage and business model innovation: The Case of the Uffizi Gallery in Florence. *Il capitale culturale. Studies on the Value of Cultural Heritage, 14*, 945–970.

Mandosi, M. (2017). Musei accessibili, musei possibili. In Cetorelli, G., Guido, M. R. (Ed.), *Il patrimonio culturale per tutti. Fruibilità, riconoscibilità, accessibilità, Quaderni della valorizzazione, 4* (pp. 47–52). Rome: Direzione generale Musei.

Marani, P. C. & Pavoni, R. (2006). *Musei. Trasformazioni di un'istituzione dall'età moderna al contemporaneo*. Venice, IT: Marsilio.

Maslow, A. H. (1943). A theory of human motivation. *Psychological Review, 50*, 370–396. 10.1037/h0054346.

Milone, V. (2004). *Il museo da entità statica a istituzione dinamica*. Bari, IT: Cacucci.

Monaci, S. (2005). *Il futuro nel museo. Il territorio e la comunicazione museale*. Milan, IT: Guerini e Associati.

Morgan, M., Lugosi, P. & Ritchie, J. R. B. (2010). *The tourism and leisure experience: Consumer and managerial perspectives*. Bristol, UK: Channel View Publications.

Pine, J. B. & Gilmore, J. H. (1999). *The experience economy: Work is Theatre and Every Business a Stage*. Boston, MA: Harvard Business Scholl Press.

Riva, G., Mantovani, F., Capideville, C., Preziosa, A., Morganti, F., Villani, D., Gaggioli, A. & Alcañiz Raya, M. (2007). Affective Interactions Using Virtual Reality: The Link between Presence and Emotions. *Cyberpsychology & Behavior, 10*(1), 45–56. doi:10.1089/cpb.2006.9993

Scartabelli, T. (2014). *Media e didattica museale nuove tecnologie per educare: quando la multimedialità ed interattività incontrano il museo e il patrimonio culturale*. Rome, IT: Lulu.com.

Schmitt, B. H. & Ferraresi, M. (2006). Marketing Esperienzale. Come sviluppare l'esperienza di consumo. *Franco Angeli Edizione*.

Seligman, M. E. P. (2011). *Flourish: A visionary new understanding of happiness and well-being*. London, UK.

Part III

Conclusion

13 Reshaping Tourism

Advances and Open Issues

Maria Della Lucia, Ernestina Giudici, and Davide Secchi

1 Opening Remarks

Humanistic Tourism offers a practical application to humanistic management, thus striving to establish a new disciplinary perspective. Most intellectual endeavors take those who engage with them to explore different, perhaps new, domains in a particular area. The idea of an intellectual 'journey' is a powerful metaphor for tourism and humanistic tourism, in particular. In the spirit of a transformational journey, in this chapter, we attempt to answer two questions, concluding with a few remarks intended as an invitation for the reader to continue the journey. The two questions, in order, are: Where has this conceptual travel taken us? And, most importantly, where should we go next?

2 Humanistic Tourism: The Journey Behind it

By explicitly marking a humanistic approach, humanistic tourism has both a critical and a positive approach to tourism management. On the one hand, it takes established perspectives in tourism management, mainly inherited from economics, human resource management, and efficiency-driven management, outlining their inadequacies. On the other hand, it discusses human dignity and values directly, thus explaining, analyzing, and interpreting the importance of topics that have been given limited attention so far. The contemporary relevance of human dignity may interpret and manage the weaknesses of traditional approaches to tourism and cope with challenges as well as new scenarios, including the current COVID-19 pandemic crisis.

These two dimensions of humanistic management can be referred to as the *humanistic difference* and the *humanistic advantage* – or the *why* and *how* of humanistic tourism. Of course, it is difficult to clearly delimit one from the other; however, a *humanistic difference* becomes apparent when humanistic tourism questions an approach that, by overemphasizing efficiency, costs, or profit, may lead to possible distortions in the supply and use of touristic services. This is by no means an appeal to abandon a healthy balance in resource management. On the contrary, it is a way to inspire the criteria through which these resources should be considered.

An example may clarify this point. At the core of the theory of choice is the concept of rationality. In his work, Herbert Simon (1997[1947]; 1955) described rationality as the selection of alternatives through a set of values, where consequences of behavior are evaluated. The picture is rather dry, and it does not really do justice to all those instances in which ambiguity, interpretation, and various limits (of the self and of the environment) come into play. Among many, scholars of sensemaking (e.g., most noticeably, Weick, 1993; Gioia, Mehra, & Weick, 1996) have highlighted these concerns regarding the classic literature on the decision making process. This should also not be intended as a call to abandon the traditional approach to decision making, in fact, some have proposed extending it (Secchi, 2011). There are many ways in which such an approach can be extended.

The one that the humanistic tourism approach adopts and proposes is that of installing a more values-based and humanistic set of norms to drive one's decision-making and actions. When this is applied to the rational choice theory, as outlined by Simon, consequences are mapped on a completely different setting. For example, if the driving criterion used to decide whether to use audio guide devices in place of human guides is simply an economic one, then the cost of a certain number of audio guide devices is compared to the costs of human guides (i.e. personnel). The outcome can be evaluated based on the higher number of customers the organization may be able to serve when using the audio guide devices, and an app, as opposed to a human guide. Of course, one usually also factors-in the quality of the service offered, but quantity tends to prevail over quality in these cases. In the decision-making process, the nature of the touristic attraction may also play a role. This simplistic example is used to show that a change of perspective – that is a change in the criterion used – causes a difference is the assessment of the consequences. When attention is given to the conveyance of meaning and, for example, to the importance of leaving a pleasant memory, providing emotional fulfilment plays in the experience, and the purely utilitarian criterion takes a "back seat". This is so if one wants to play it, remaining within the domain of the decision making literature, closer to a logic of *appropriateness* rather than to one of *consequences* (March & Heath, 1994). Furthermore, what is appropriate has to engage with identity, dignity, and a responsible, as well as sustainable, use of resources.

The second perspective, that of a *humanistic advantage*, is, instead, dedicated to developing a way to study and manage tourism that includes humanistic education. Considering this, the objective is to unveil and demonstrate how many aspects of tourism can emerge by looking at a values-driven humanistic perspective. Continuing with the decision making example above, this is equivalent to taking a sensemaking perspective. By paying attention to how individuals use their mind frames to interpret cues from the environment and from others to give meaning to what is happening, they make sense of situations (e.g., Weick, 1993). This approach has been a game changer, especially if one considers the rather emotionless perspective outlined in the paragraph above. In this respect, there is no need to refer to the traditional way of doing things. The intent is different: Explaining understanding, and showcasing what would be difficult otherwise. The *humanistic advantage* takes this unconstrained perspective to the study and the management of tourism.

3 The Journey Ahead: Where Next?

The *why* and *how* of the humanistic tourism journey have shown that awareness of *dignity* is a foundational element to inform appropriate approaches to tourism, as is the importance of *values and norms* when addressing contemporary challenges in tourism. Finally, it highlights the value of humanistic *education*, which is viewed as indispensable to keep tourism practice at pace with rapidly evolving environments.

These dimensions of humanistic tourism's difference and advantage recall the interdependences – for example between tourism theory, practice, and education – needed to make a more values-based and humanistic approach to tourism effective. The interpretation of tourism organizations and their environments as complex systems may be a possible way forward. By so doing, the concepts of "systems" and "complexity" are briefly presented and then applied to tourism.

3.1 Systems and Systems Thinking

The idea of considering scientific phenomena as systems was promoted long ago by von Bertalanffy (2015[1968]) and a community of scholars that intended the concept as a way to make knowledge easier to transfer (Heims, 1991; Ramage & Shipp, 2020). This is simply to state that the idea is not new and that it has been entrenched and intertwined with scientific discourse across disciplines for quite some time. It implies that the concept of "system" has become part of the way in which some disciplines treat their core concepts, including management and organization research.

The idea of a system is rather simple, it is something composed by *parts* whose *interconnections* define its *functionality* and, at the same time, establish relations with the *outside* world. There are four elements in this definition, which loosely track the way in which W. R. Scott (2002) uses the concept in his book on organizations as systems. First, it is stated that a system is composed of different parts. This identifies a very large number of observable and non-observable items, because almost everything can be divided into parts. However, simply dividing is not the point here. These parts need to be organized in a way such that their interconnections, their exchanges, constitute what matters. This is the second element of the definition. In other words, it is the organization of parts that matters in a system. It is not the simple juxtaposition we are discussing here; it is the way in which the parts relate to each other that matters. In light of this, the fact that tourist organizations (and destinations) can be divided into various types of elements is not all that interesting. What makes us explain and understand what such an organization does is the way in which its participants utilize the material and immaterial resources available. But this means that we study tourism organizations as systemic entities. Once again, interrelations matter!

The third element of the definition is *functionality*. This is to be intended more in terms of its operations rather than an aim or objective that the system has. In fact, whether a system needs an ultimate goal to be called as such is still an open question today, and this is not the place to take this question on (see Cowley & Vallée-Tourangeau, 2017, for not-strictly functionalistic approaches applied to systemic cognition). The definition simply states, as a matter of fact, that the way in which parts are connected to each other gives an idea of how the system operates as such. A simple example of this is the hierarchical chain of command in a touristic organization, or the coordination mechanisms used in a destination management organization (Beritelli, Bieger, & Laesser, 2007). This element affects the way in which operations are conducted and contributes to identifying the system-organization. Top-down governance, shared governance and intelligent governance do identify different destination management models (Trunfio & Della Lucia, 2019), which capture the shift toward system planning and integrated development (Hall, 1998) prompted by socioeconomic and technological changes.

Lastly, comes what we referred to as the *outside* world. This is usually called the *environment*, and is defined by difference: what is not part of the system must be outside of it. This does not mean that it is irrelevant, quite the opposite. Depending on how one defines it, the environment has a fundamental role in determining how the relations within the system are shaped. There are multiple examples, especially from an organizational theory point of view. Institutional theory, for example, posits that any organization is challenged by pressures to conform that come from a multitude of sources, some direct (e.g., legislation), others indirect (e.g., professional associations; Powell & DiMaggio, 1991; Greenwood, Raynard, Kodeih, Micelotta, & Lounsbury, 2011). This is, again, a definitional aspect: the system cannot exist in a vacuum.

In more recent years, management scholars have considered the above and created an approach to the study of organizations. This was labeled *systems thinking* (e.g., Checkland, 1999; Jackson, 2002), and is an attempt to pay attention to interconnections without missing the overall picture, while also acknowledging the importance of relations with the (various) environment(s).

3.2 Complexity and Adaptive Systems

Complexity can and has been used in several ways to define business organizations, and organizations in general, as adaptive systems. In other words, the idea of complexity is related to that of systems and, at the same time, to the idea that any system *changes* over time (Burnes, 2005). This concept of evolution, dynamics, adaptation, or simply movement from one state to another, is very much embedded in any organization (Goldstein, Hazy, & Lichtenstein, 2011). Hence, *time* is a key component in the study of any system. Organizations, and tourism organizations, are no different.

When an organizational system is considered in its dynamic change over time – that is, we assume that it is not static – there are several additional elements that come to the fore. Once again, this is not the right place for an in-depth presentation and analysis of these elements but, at the same time, we can sketch those that seem more relevant to our case. They are *uncertainty*, *heterogeneity*, and *context*, and they identify organizations as complex systems.

3.2.1 Uncertainty

There are multiple ways in which an organization can change. This is because there are multiple sources of pressure and none of them usually has a univocal way of manifesting itself. Employee input, changes in legislation, political pressures, climate change, and pandemic crises, all need to be interpreted and ultimately balanced. As an organization strives to balance all these factors, it necessarily deals with uncertainty – that is the lack of knowledge on how probable events are. As mentioned, these uncertainties are sometimes difficult to interpret and, therefore, ambiguous, non-linear, and messy (Abrahamson, 2002; Herath, 2019). The implications of this reality are many. Perhaps the most relevant are that: (a) an organization needs to adapt constantly; and (b) the ways of adaptation are unpredictable.

3.2.2 Heterogeneity

Not only there is a mix of internal and external sources of pressure, but the factors/actors that operate within and outside the organization can be very different. For example, stakeholders – not only employees, managers, executives, but also customers, lawmakers – are different, and differences span both within and between each category. It is almost impossible to predict the behavior of an organization when reasoning on the average employee, for example. Understanding and analysis are necessarily tied to the extent to which differences and variety are both considered, valued, and managed (Beer, 1970). A complex system should be considered in its heterogeneous components.

3.2.3 Context

The various components of an organization only make sense when situated in their local contexts. For example, the patterns of employee motivation, job satisfaction, and perceptions of fairness are affected by interactions with those with whom they are in contact the most (Secchi & Gullekson, 2015). Clearly, one could also observe a general organizational effect, but that is more an overall conditioning factor that serves as a constraint for more localized and contextual dynamics. Especially in large organizations, the exercise of describing an overall balance (equilibrium) is futile, unless it is matched with the various multiple balances (or, sometimes, imbalances) that appear in the various "local" contexts.

The three elements above describe organizational change through time as unpredictable, ambiguous, anchored to heterogeneous agents, and bound to local dynamics. These are some of the characteristics of a complex system that adapts to internal and external stimuli.

4 A Complexity Perspective for Humanistic Tourism

The extension to humanistic tourism of a complex adaptive systems perspective (e.g. Fioretti, 2016) allows the interpretation of (the interplay between) dignity, values, and education through the lenses of uncertainty, heterogeneity, and context. It also paves the way to further developing humanistic tourism through the connections that tourism organizations have with their environment.

The study of complexity is usually conducted on the various levels in which parts of a system interact, hence moving from a micro/individual to a 'meso domain' (Secchi & Cowley, 2018) and to a macro-structural domain (where patterns may emerge; see Grimm et al., 2005). From this approach, there are trickle-down effects that force the reconsidering of the role of both individuals and of the various meso and macro domains. Hence, the complexity perspective has implications for the role of the individual, and for a way to establish a macro perspective through a sustainability route. If taken to their core message, these implications enable both the use of dignity and values-driven approaches in the study and management of tourism and education in tourism management.

4.1 The Individual in the Tourism Organization

From the humanistic tourism perspective, the role of the individual is associated with the discourse on *dignity* – that is, the recognition and protection of human value – which, in turn, is intertwined with one of identity and respect. By acknowledging that tourism organizations are complex (Goldstein et al., 2011), one also necessarily recognizes that every part needs to be valued with respect to its own identity, as opposed to that of the organization as a whole. Under circumstances of injustice and unfairness, that is the failure to acknowledge human dignity, there is a failure in the understanding of the diversity of the entities in a system. Of course, acknowledgement is not a neutral stance, it is a *values* stance. In fact, it comes with the recognition of the value and the respect that persons, objects, as well as environments, deserve. In turn, by acknowledging the fact that organizational stakeholders matter and have value, as long as they are considered in their heterogeneity and diversity, the complexity perspective acknowledges they are tied to local conditions, situations, histories, and cultures.

If viewed under this lens, individual agents should be not considered as mere executors, as passive, or as easily influenced by exogenous forces. Nor can we consider them as completely isolated from each other. In other words, we should re-think them as having a rich cognitive backbone that allows plastic adaptation, as well as autonomy (Secchi, 2019). Their exchanges are not a mere transfer of data or information, but also include their emotionality, identities, uncertainties, culture, beliefs, etc. This means that most of what agents are depends on and results from their sociality: how they interact with others, with ideas, and with the artifacts around them (Hollan, Hutchins, & Kirsh, 2000).

From a cognitive perspective, individual resources are necessarily distributed among people, things, and ideas (Hutchins, 1995). Put differently, an individual is never an isolated monad, but is a systemic interactive element (Secchi, 2011). If we embrace the systemic/distributed cognitive perspective above the reason is apparent, and it is that the individual is ecologically embedded in the system (see Varela, Thompson, & Rosch, 1991). A failure to recognize this is a failure to acknowledge the rich substratum of what makes us human.

In terms of humanistic tourism, a complex perspective means that every individual must be holistically recognized and valued, as well as considered as part of a system of interrelations, each unique, in some way, and each potentially worthy of preservation. Diversity/heterogeneity and context are intrinsic features of tourism systems, as this is a labor-intensive industry, employing informal workers, minorities, women, migrants, and young people. It is also a global industry, allowing and fostering encounters between people, languages, cultures, and traditions in different socio-cultural contexts. Sociality is also a feature of an information-intensive industry and an "experience good", resulting from the relationship and co-creation processes involving hosts and guests in different places (Gretzel, Fesenmaier, Formica, & O'Leary, 2006). Experiential tourism and creative tourism do extract value from diversity, uniqueness, and authenticity (Tiberghien, Bremner, & Milne, 2020) and through "meeting locals and living like a local" (Richards, 2014).

4.2 The Individual and the Tourism Ecosystem

The resources around individuals may be interpreted in the same perspective, this time by taking the angle of a macro-structural domain. We have been made abundantly aware that the clock of life, as we know it on this planet, has started to tick a little faster. The proverbial scarce resources of economic memory may become so limited so fast that action is urgently required. The consumption of unreproducible resources goes along with social issues including decent work, equity, justice, fairness, and social interactions.

The call to action comes from, among many, the United Nations' *Sustainable Development Goals* (UN, 2015). Furthermore, "Dignity, People, Prosperity, Social Justice, Planet, and Partnerships" have been identified as essential requirements for their achievement. The issues are of such magnitude that there are few simple solutions. At the same time, every single one of us and our organized lives are called to coordinated action.

Tourism is a set of activities that significantly affects sustainable development – tourist transport system among others (Scuttari, Della Lucia, & Martini 2013) – and is included in the actions for sustainable development (Della Lucia, 2018). The recognition of heterogeneity, at both the micro and meso levels, enables various parts to be valued for the value they have and the way in which they interact (with others, as well as with ideas and artifacts).

In an ecological view of individual cognition (Secchi, 2009) – that is the distribution of individual cognitive resources among people, things, and ideas mentioned above – interrelations are made to work and there is an inherent tendency to preserve them, both within and outside the organization. In a multi-faceted approach to the individual agent, the likely outcomes to their negotiations, cooperation, altruism, and (social) responsibility depend on how individual agents are. This, in turn, is a result of their sociality. In this line of intentionality, the individual wants to act responsibly, in order to be able to use the same resource in the future. However, in recent years, the literature on corporate social responsibility has turned to the study of the individual (e.g., Glavas, 2016; Rupp, Skarlicki, & Shao, 2013), recognizing and valorizing the role of people as core components of any socially responsible program.

In this stream of literature, the socially responsible individual is necessarily one who cares about a sustainable future (e.g., Onkila, 2017; Pearce & Stahl, 2015; Rupp & Mallory, 2015).

This turn in the multi-faceted approach to the individual agent goes in parallel with the turn from reactive to strategic-oriented approaches to business and social responsibility in tourism (Farmaki, 2019). In the former, the internalization of natural and cultural resources (and related agents) in the firm's strategies and operations is aimed at enhancing its competitive advantage and at minimizing the negative effects of tourism business activities. In strategic-oriented approaches, responsibility is integrated also in the firm's value chain, enabling companies and society to mutually benefit by reconciling different stakeholder interests. This approach links individual responsibility to corporate responsibility to sustainability through the fundamental tenets of sustainable (tourism) development (Bramwell & Lane, 2011). Holistic planning and stakeholder participation frame tourism within the wider process of territorial governance and allow space for the broad-based, equitable decision-making processes (Della Lucia & Franch, 2017). Shared governance (the sharing of authority and responsibility) – often referred to as co-management, collaborative management, joint management, or multi-stakeholder management – is becoming more common in many complex systems. In fact, individual agents actively participate in this system of interrelations. Therefore, it is not a stretch to claim that this future orientation is exactly what socially responsible behavior looks like, both within and outside organizations. In other words, sustainability is the most likely outcome of a complex approach to organizations and individuals.

4.3 Humanistic Tourism Education

Education is another dimension of humanistic tourism's differences and advantages. When we frame organizations as complex adaptive systems, we are stating that time and change are particularly relevant for them. A feature of organizations is that their behavior is unpredictable due to uncertainty; an organization needs to adapt constantly, and the ways of adaptation are unpredictable. This implies that: (a) there is only a loose connection, if at all, with the conditions in which a process has started; and (b) many behaviors emerge as the system moves on. If education is a way to improve and innovate tourism, for example humanistic tourism, then it becomes crucial that the origins of change (concisely summarized in the two points above) are fully acknowledged.

The first feature (a), is also called independence from initial conditions (Prigogine & Stengers, 1979), and it indicates that despite even complete knowledge of the conditions in which an organization is today, it is very difficult to predict where this same organization will be in a future time.

The second feature (b) is a property of the system. When behaviors or patterns, in general, appear to be difficult to connect to the conditions in which the system was in the past, it is said the system has shown *emergent* properties (Cunningham, 2001). These emergent properties of a system need to be monitored very closely, especially when they appear in an organization. This is because, when they bear positive results, managers may want to know how to duplicate them. Conversely, if the results are negative, they would want to eliminate them. The value propositions of humanistic management (Melé, 2016) and, in turn, of humanist tourism, may be considered emergent properties of tourism when they develop from the related business activities, independent of the underlying assumptions of business operations. The development of humanistic management studies is needed to shed light on emerging values-based and humanistic sets of norms and patterns.

Being not easy to control and manage, complex systems require quite significant investments in knowledge, the constant collection of data and observations, as well as analyses. To enable a dignity- and values-driven approach to study and manage tourism in this perspective, a focus on education in emerging properties is needed. This is related to the intricacies of change. If, as outlined above by the two points on the dependence of initial conditions (a) and emergent properties (b), change means an inherent difficulty in understanding and outlining cause-effect mechanisms, then attention to education becomes central (Giudici, Dettori, & Caboni, 2020). Furthermore, this attention is to be renewed constantly, in line with the continuous adaptation of values, systems, people, and the environment.

5 Concluding Remarks

The humanistic tourism journey reveals two perspectives on tourism, one is the *humanistic difference* and the other is the *humanistic advantage*. The two perspectives allow the extending to humanistic management of a complex systems approach, thus reinterpreting human dignity, values, norms, and education through the lenses of uncertainty, heterogeneity, and context(s). This approach also paves the way for further developing humanistic management. The trajectories are two, among others.

The first stream of research entails a specific area of enquiry that naturally connects to tourism as a complex system. In the wake of global challenges, including climate change and pandemic crises, social responsibility and sustainability have become prominent concepts and practices in tourism. At their core, there is the need to multiply the nature (economic, environmental, and social) and sources of value generation. In turn, it recalls the ecosystem of value co-creation (and co-destruction) (Cabiddu, Lui, & Piccoli, 2013), mobilizing resources and capabilities from many entities, sharing the costs of social transformation and the new economic opportunities arising from social progress. In the background there are synergies, enabling mechanisms and constraints that may derive by seemingly independent elements of these systems. Most of the causal mechanisms underneath these processes of value co-creation and co-destruction are still unclear because tourism is a complex system. When studying sustainability and social responsibility in the humanistic tourism perspective, complexity should be kept in mind. Individual agents actively participate in systems of interrelations as complex human beings, recognizing their interconnection with other people, the environment, and all living things (Melé, 2016). Their search for both freedom and common welfare may create tensions between egoism and responsibility.

The second stream of research is strictly connected and has a methodological nature. The values-based humanism and humanistic perspectives to tourism require non-reductionism as well as a true systemic approach. The intent is not to deny the progress that science has achieved through incremental reductionism, but to highlight the further increasing advances that are possible by taking a different perspective. Qualitative research is generally in line with the claims made here, although some researchers may want to push the boundary further. A complex systems perspective would urge researchers to consider *social* units of analysis, and the meso domain of social interactions (Secchi & Cowley, 2018), meant as wider integrations and combinations of individuals as a basis for their enquiries. This requires not only a wider set of qualitative tools (e.g., video-analysis, ethnography, conversation analysis, cognitive event analysis), but it opens the way to agent-based computational simulations (e.g., Secchi, 2015; Secchi & Neumann, 2016) as a way to bridge the gap.

References

Abrahamson, E., (2002). Disorganization Theory and Disorganizational Behavior: Towards an Etiology of Messes. *Research in Organizational Behavior*, *24*, 139–180. doi: 10.1016/s0191-3085(02)24005-8.

Beer, S., (1970). Managing Modern Complexity. *Futures*, 2(3), 245–257. doi: 10.1016/0016-3287(70)90028-5.

Beritelli, P., Bieger, T. & Laesser, C., (2007). Destination Governance: Using Corporate Governance Theories as a Foundation for Effective Destination Management. *Journal of Travel Research*, *46*(1), 96–107.

Bertalanffy, L. V., (2015). *General system theory: Foundations, development, applications*. New York: George Braziller.

Bramwell, B. & Lane, B., (2011). Critical Research on the Governance of Tourism and Sustainability. *Journal of Sustainable Tourism*, *19*(4–5), 411–421.

Burnes, B., (2005). Complexity Theories and Organizational Change. *International Journal of Management Reviews*, *7*(2), 73–90. doi: 10.1111/j.1468-2370.2005.00107.x.

Cabiddu, F., Lui, T. & Piccoli, G., (2013). Managing Value Co-creation in the Tourism Industry. *Annals of Tourism Research*, *42*, 86–107.
Checkland, P., (1999). *Systems Thinking, Systems Practice*. Hoboken: John Wiley & Sons.
Chuang, S. T., (2013). Residents' Attitudes Toward Rural Tourism in Taiwan: A Comparative Viewpoint. *International Journal of Tourism Research*, *15*(2), 152–170.
Cowley, S. J. & Vallée-Tourangeau, F., (2017). *Cognition Beyond the Brain*. Cham, Switzerland: Springer International Publishing.
Cunningham, B., (2001). The Reemergence of 'Emergence'. *Philosophy of Science*, *68*(S3), S62–S75. doi: 10.1086/392898.
Della Lucia, M., (2018). *Approccio manageriale allo sviluppo turistico sostenibile. Un framework interpretativo e di gestione integrato*. Milano: Franco Angeli.
Della Lucia, M. & Franch, M. (2017). The Effects of Local Context on World Heritage Site Management: The Dolomites Natural World Heritage Site, Italy. *Journal of Sustainable Tourism*, *25*(12), 1756–1775. doi: 10.1080/09669582.2017.1316727.
Fioretti, G., (2016). Emergent Organizations. In Secchi D. & Neumann M. (Eds.), *Agent-Based Simulation of Organizational Behavior* (pp.19–41). doi: 10.1007/978-3-319-18153-0_2
Gioia, D. A., Mehra, A. & Weick, K. E., (1996). Sensemaking in Organizations. *The Academy of Management Review*, *21*(4), 1226. doi: 10.2307/259169.
Giudici, E., Dettori, A. & Caboni, F., (2020). Challenges of Humanistic Management Education in the Digital Era. In R. Aguado & A. Eizaguirre (Eds.), *Virtuous Cycles in Humanistic Management. From the Classroom to the Corporation* (pp. 21–35). Switzerland: Springer Nature.
Glavas, A., (2016). Corporate Social Responsibility and Organizational Psychology: An Integrative Review. *Frontiers in Psychology*, *7*(144), 1–13. doi: 10.3389/fpsyg.2016.00144.
Goldstein, J. H., Hazy, J. K. & Lichtenstein, B. B., (2011). *Complexity and the Nexus of Leadership: Leveraging Nonlinear Science to Create Ecologies of Innovation*. New York: Palgrave Macmillan.
Greenwood, R., Raynard, M., Kodeih, F., Micelotta, E. R. & Lounsbury, M., (2011). Institutional Complexity and Organizational Responses. *The Academy of Management Annals*, *5*(1), 317–371. doi: 10.1080/19416520.2011.590299.
Gretzel, U., Fesenmaier, D. R., Formica, S. & O'Leary, J. T., (2006). Searching for the Future: Challenges Faced by Destination Marketing Organizations. *Journal of Travel Research*, *45*(4), 116–126.
Grimm, V., Revilla, E., Berger, U., Jeltsch, F., Mooij, W. M., Railsback, S. F., … & DeAngelis, D. L., (2005). Pattern-Oriented Modeling of Agent-Based Complex Systems: Lessons from Ecology. *Science*, *310*(5750), 987–991. doi: 10.1126/science.1116681.
Hall, C., (1998). *Introduction to Tourism: Development, Dimensions and Issues* (3rd ed.). Melbourne: Longman.
Heims, S. J., (1991). *The Cybernetics Group*. Boston: The MIT Press.
Herath, D. B., (2019). *Business Plasticity Through Disorganization*. Bingley, UK: Emerald Publishing.

Hollan, J., Hutchins, E. & Kirsh, D., (2000). Distributed Cognition. *ACM Transactions on Computer-Human Interaction (TOCHI)*, *7*(2), 174–196. doi: 10.1145/353485.353487.

Hutchins, E., (1995). *Cognition in the Wild*. Boston: MIT Press.

Inskeep, E., (1991). *Tourism Planning*. New York, NY: Van Nostrand Reinhold.

Jackson, M. C., (2002). *Creative Systems Thinking: A Complete Approach for Successful Management*. Hoboken: Wiley.

March, J. G. & Heath, C., (1994). *A Primer Decision Making: How Decisions Happen*. New York: Free Press.

Melé, D., (2016). Understanding Humanistic Management. *Humanistic Management Journal*, *1*(1), 33–55. doi: 10.1007/s41463-016-0011-5.

Onkila, T., (2017). Employee Rhetoric in the Acceptance or Rejection of Corporate Environmentalism. *Organization & Environment*, *30*(2), 142–161.

Pearce, C. L. & Stahl, G. K., (2015). Introduction to the Special issue: The Leadership Imperative for Sustainability and Corporate Social Responsibility. *Organizational Dynamics*, *44*, 83–86.

Powell, W. W. & DiMaggio, P., (1991). *The New Institutionalism in Organizational Analysis*. Chicago: University of Chicago Press.

Prigogine, I. & Stengers, I., (1979). *La nouvelle alliance. Métamorphose de la science*. Paris: Gallimard.

Ramage, M. & Shipp, K., (2020). *Systems Thinkers*. Berlin: Springer.

Richards, G., (2014). Creativity and Tourism in the City. *Current Issue in Tourism*, *17*(2), 119–144.

Rupp, D. E., Skarlicki, D. & Shao, R., (2013). The Psychology of Corporate Social Responsibility and Humanitarian Work: A Person-Centric Perspective. *Industrial and Organizational Psychology*, *6*(4), 361–368. doi: 10.1111/iops.12068.

Rupp, D. E. & Mallory, D. B., (2015). Corporate Social Responsibility: Psychological, Person-Centric, and Progressing. *Annual Review of Organizational Psychology and Organizational Behavior*, *2*(1), 211–236.

Scott, W. R., (2002). *Organizations: Rational, Natural and Open Systems*. Upper Saddle River, New Jersey: Prentice Hall.

Scuttari, A., Della Lucia, M. & Martini, U., (2013). Integrated Planning for Sustainable Tourism and Mobility. A Tourism Traffic Analysis in Italy's South Tyrol Region. *Journal of Sustainable Tourism*, *21*(4), 614–637.

Secchi, D. & Neumann, M., (2016). *Agent-Based Simulation of Organizational Behavior: New Frontiers of Social Science Research*. Cham, Switzerland: Springer International Publishing.

Secchi, D., (2009). The Cognitive Side of Social Responsibility. *Journal of Business Ethics*, *88*(3), 565–581. doi: 10.1007/s10551-009-0124-y.

Secchi, D., (2011). *Extendable Rationality: Understanding Decision Making in Organizations*. Berlin: Springer.

Secchi, D. & Gullekson, N. L., (2015). Individual and Organizational Conditions for the Emergence and Evolution of Bandwagons. *Computational and Mathematical Organization Theory*, *22*(1), 88–133. doi: 10.1007/s10588-015-9199-4.

Secchi, D. & Cowley, S. J., (2018). Modeling Organizational Cognition: The Case of Impact Factor. *Journal of Artificial Societies and Social Simulation*, *21*(1), 1–13. doi: 10.18564/jasss.3628.

Secchi, D., (2015). A Case for Agent-Based Models in Organizational Behavior and Team Research. *Team Performance Management: An International Journal*, *21*(1/2), 37–50. doi: 10.1108/tpm-12-2014-0063.
Secchi, D., (2019). Foreword. In Herath, D. (Ed.), *Business Plasticity Through Disorganization (xv-xxiii)*. Bingley, UK: Emerald Publishing.
Simon, H. A., (1997). *Administrative Behavior: A Study of Decision-Making Processes in Administrative Organization*. New York: Free Press.
Simon, H. A., (1955). A Behavioral Model of Rational Choice. *The Quarterly Journal of Economics*, *69*(1), 99. doi: 10.2307/1884852.
Tiberghien, G., Bremner, H. & Milne, S., (2020). Authenticity and Disorientation in the Tourism Experience. *Journal of Outdoor Recreation and Tourism*, *30*, 1–9. doi: 10.1016/j.jort.2020.100283.
Trunfio, M. & Della Lucia, M., (2019). Engaging Destination Stakeholders in the Digital Era: The Best Practice of Italian Regional DMOs. *Journal of Hospitality & Tourism Research*, *43*(3), 349–373. doi: 10.1177/1096348018807293.
United Nations (UN). (2015). Transforming Our World: The 2030 Agenda for Sustainable Development. Retrieved from https://sustainabledevelopment.un.org/content/documents/21252030%20Agenda%20for%20Sustainable%20Development%20web.pdf.
Varela, F. J., Thompson, E. & Rosch, E., (1991). *The Embodied Mind: Cognitive Science and Human Experience*. Boston: MIT press.
Weick, K. E., (1993). The Collapse of Sensemaking in Organizations: The Mann Gulch Disaster. *Administrative Science Quarterly*, *38*(4), 628. doi: 10.2307/2393339.

Index

Index Book I - Humanistic tourism: Values, Norms and Dignity (ed. by Maria Della Lucia and Ernestina Giudici)

University museum system 196, **198**, **200–01**, 205

For Product Safety Concerns and Information please contact our EU
representative GPSR@taylorandfrancis.com
Taylor & Francis Verlag GmbH, Kaufingerstraße 24, 80331 München, Germany

www.ingramcontent.com/pod-product-compliance
Lightning Source LLC
LaVergne TN
LVHW021121300926
849160LV00010B/814

* 9 7 8 0 3 6 7 6 2 3 4 2 5 *